The Signpost of God

THE SUBLIME SECRET OF THE HOLY GRAIL

-

FIRST EDITION

Douglas James Meyer

Aletheia Press

Middletown, New Jersey, USA

Douglas J. Meyer

Middletown, New Jersey, USA

www.douglasjamesmeyer.com

Book Layout ©2017 BookDesignTemplates.com. Image of Tudor Dynasty Rose reproduced with license to CrystalGraphics.com. All other images are rendered by the author. The Orrery program is written by the author, algorithms and equations and VSOP87 planetary tables are from, *Astronomical Algorithms, Second Edition* by Jean Meeus (1998), and are reproduced with permission from the publisher Willmann-Bell Inc., Richmond Virginia.

The Signpost of God / Douglas J. Meyer. —1st ed. AMZv1.618

ISBN 978-1-7322721-2-5

The Keystone of the Grail Mystery
is Matthew 16:18

The Holy Grail is a multi-faceted Gem...

For Scientists:
It is an astronomical phenomena
that may lead humanity
to a new Earth around a distant star.

For Historians:
It is the only rational explanation of
the medieval persecution of astronomers
and a sun-centered cosmology.

For Theologians:
It is the "stumbling stone" upon which
they tripped and fell, but which may
become; the "headstone of the corner"
and "marvelous in our eyes".

Contents

Introduction

For the Master, the Compass of Faith is above the Square of Reason; but both rest upon the Holy Scriptures and combine to form the Blazing Star of Truth.
Morals and Dogma - Albert Pike 1871

It is a common saying that if you wish to hide something of great value, you should leave it out in the open. Such it is with the Holy Grail, which has been a great unsolved mystery for 800 years. However, nothing short of divine providence has willed it to be in the open for all to see as an astronomically large Signpost of God. The thesis of this book is that the Grail is related to an astronomical phenomenon mystically connected to the genesis of life on Earth – and for this reason it is called Holy. Fully understanding this particular Grail secret is very significant for understanding the medieval clerical antipathy toward astronomy. In fact, this is the most compelling explanation of why the Medieval Church condemned a sun-centered cosmology as a heresy. You could never demonstrate the Grail secret as long as it was illegal, and punished with death, to depict the planets in orbit around the sun. Rather than the Grail being some physical artifact to be found and placed in a museum or lost forever, the Grail is eternal and accessible to all the people in the world. To exactly reproduce this astronomical phenomena, the Appendix of this book contains computer programming code based on, "Astronomical Algorithms" by Jean Meeus.[1]

In the story of the Garden of Eden, God commands Adam and Eve that they shall not eat of the fruit of the tree of the knowledge of good and evil. The serpent told Eve that if she ate the forbidden fruit, knowledge would open her eyes to good and evil. Eve tempts Adam to eat the fruit. After he eats from the tree of the knowledge of good and evil we read, "Behold, the man has become like one of us, knowing good and evil; and now lest he put forth his hand and take also from the tree of life [...] therefore the Lord God sent him forth from the Garden of Eden, to till the ground from whence he came." (Gen 3:22-24) In words attributed to Christ, *he associates himself with the serpent in Eden offering forbidden fruit*. Jesus says, "To he that overcomes, I will give to eat of the tree of life, which is in the paradise of God." (Rev 2:7) In his book "A Practical Guide to Qabalistic Symbolism," Gareth Knight wrote the following about the serpent in Eden.

> Let the disciple seize hold of the tail of the serpent of wisdom, and having with firmness grasped it; let him follow it into the deepest center of the Hall of Wisdom.[2]

Ministers of Christ are supposed to be stewards of these mysteries to which the scriptures bear witness, "But we speak the wisdom of God in a mystery, even the hidden

---

[1] Jean Meeus, *Astronomical Algorithms, Second Edition*. (Richmond Virginia: Willmann-Bell, Inc., 1991, 1998).

[2] Gareth Knight, *A Practical Guide to Qabalistic Symbolism*. (York Beach Maine: Samuel Weiser, Inc. 1965). Vol.2 p.259

[occult] wisdom, which God ordained before the world unto our glory." (1 Cor 2:7) Christ is said to have fulfilled Old Testament prophecy, in that he revealed ancient knowledge, "I will utter hidden things, things from of old." (Ps 78:2) It is written that Jesus fulfilled prophecy where the prophet said, "I will open my mouth in parables, and I will utter things which have been kept secret from the beginning of the world." (Matt 13:35)

One secret that has been kept from the beginning of the world is that of the Grail dating back to the origins of modern civilization at least as early as 4,000-3,200 BC. Knowledge of the Grail mystery is timelier now than ever in the past, because it could promote a new science that may lead humankind to a destiny in the Cosmos. Few people fully understand the significance of the quote preceding this introduction by Albert Pike, the famous nineteenth century Freemason, "For the Master, the Compass of Faith is above the Square of Reason; but both rest upon the Holy Scriptures and combine to form the Blazing Star of Truth."[3] In this book is an explanation of exactly how Holy Scripture combined with science forms the Blazing Star of Truth and an argument that this is the gospel of forbidden knowledge of the Holy Grail. This may very well be the key to unlock the mystery of the origin of life in the Cosmos. To the best of the author's knowledge, no one has ever before published this solution of the Grail mystery. This theory of the Grail is crystalline and clear, perfect and separate from worldly matters, and yet this knowledge became forbidden and associated with Satanism, witchcraft, and the antichrist. It is because of its controversial nature that one Grail romancer indicated that while it is a delight to those who are worthy of its secret, it cannot tolerate the presence of the sinful. There

---

[3] Albert Pike, *Morals and Dogma of the Ancient and Accepted Scottish Rite of Freemasonry*. (Montana USA: Kessinger Publishing Company, 1871) p.841.

is Biblical evidence, that this scientific knowledge of the Grail was the very foundational rock upon which an original spiritual church was built, *yet it became a secret for which very many people would die.*

Part 1 - As Above

In which the foundation of the Grail symbolism is examined in Medieval Grail romances; the primitive Christian cult is speculated upon; and how the secret of the Grail was suppressed along with the Goddess, the Pentacle, and Science as Sorcery or Witchcraft.

CHAPTER 1

On This Rock

And I tell you, you are Peter,
and on this rock I will build my church,
and the powers of death shall not prevail against it.
Matt 16:18 RSV

Once upon a time, during the Dark Ages, the possession of forbidden knowledge related to religion could earn you a death sentence. There was then, and even still in modern times the concept of a heresy. This is generally an idea or opinion about religious doctrine at variance with orthodox religious teaching. To ancient theologians, heresy was like a viral infection that could infect one person and then spread among their flock like a deadly plague. This is why theologians often referred to heresy with the adjective "pestilent." The only difference between a real plague and heresy was that the suffering and death came not from nature, but from the representatives of God.

Millions of people have been intrigued by the secret mystery that is known as the "Holy Grail." For hundreds of years this enduring mystery has been a very peculiar sort of heresy. In its most common symbolic form, it is the cup of Christ used at the Last Supper, but this is just an inoculation idea. It is kind of like a flu shot containing a modified germ to stimulate a resistance to a heretical disease. Because the Grail itself is eternal and pre-dates Christianity, making it the cup of Christ was a way to convert the Grail to Christian symbolism in a time when any non-Christian ideology was persecuted or destroyed. To find the secret of the Grail and understand its mystery, one may look diligently for what could be *the greatest heresy ever told*.

In the Book of Revelation, it is recorded that a great burning star was cast out of heaven and from the theological blueprint for life, and it fell to earth and was called Wormwood. Wormwood is a plant that produces absinthe, a mind altering and possibly brain-damaging drug. This burning star could alter your mind, and damage your perception of religious teachings. Therefore, theologians would have you believe the burning star may put your soul in peril of eternal damnation. The verses from Revelation symbolically record how and why the Holy Grail and its secret were omitted from orthodox Christianity. Depending on your particular perspective, the Grail secret may be the truth that sets you free or it may cast you out of Eden for eating its forbidden fruit. The mystery of the Grail is an ancient secret for one very precise reason. It is supremely heretical because its truth has been perceived as undermining faith and ecclesiastical authority. Beware, for one may propose medieval theologians viewed the secret of the Grail as a particularly virulent heresy that would forever damn you just by exposure.

One specific heresy that was punished with the death penalty was the scientific truth about heliocentricity. This was the astronomical knowledge that placed the Sun, rather than the Earth, at the center of the solar system. The usual explanation given for outlawing this scientific truth is that it does not place the Earth and humanity as central to God's creation. This reasoning is nothing but a smoke screen or obfuscation of the truth. The persistent persecution of heliocentricity began with the political rise of Christianity in the fourth century. For about 1,200 years heliocentricity was persecuted as a heresy for which people were tortured and executed. The secret of the Grail as proposed in this book is the only specific and logical explanation for the persecution of heliocentricity. This is the smoking gun evidence by negation, that this proposed solution to the Grail mystery is perhaps a very proud and unrepentant heresy.

Matthew 16:18, refers to a rock, "And I tell you, you are Peter, and on this rock I will build my church, and the powers of death shall not prevail against it." Matthew 16:18, has previously been the foundational gospel verse upon which the Roman Catholic Church based its power and religious authority. To this end, the Church has promoted a concept of apostolic succession from Peter the rock, but the Magisterium's argument of apostolic succession is specious or only superficially plausible. Here you will learn a very different interpretation of this passage. It is a key to unlock stubborn mysteries of the Christian mythology such as explaining why December 25th was chosen as the day of the Savior's birth and January 6th the celebration of an Epiphany. For more than any other reason, Matthew 16:18 is important for precisely where it is placed in the Gospels. The fact of this placement is of central importance to this argument. It shows the work of an anonymous hand preserving the secret of the Grail. The position of the verse, together with the fact that it references a rock, joins together literary

Grail symbolism to scripture in both the Old and New Testaments.

In one of the most significant Grail romances the Grail is not symbolized by a cup, rather it is a rock or stone with special powers to bestow life and sustenance. I will demonstrate how the rock of Matthew 16:18 is a reference to the Grail and show why this interpretation is uniquely heretical. It proves that the underlying secrets of the Christian symbolism originated in the scientific teachings of the ancient pagan world. To this purpose, assuming you know nothing about the Grail we shall now begin a discussion of the primary medieval sources of the Holy Grail lore. The Grail Quest is enshrined in all the stories about Britain's King Arthur and the Knights of the Round Table. Although not prominent in all Arthurian tales, the Holy Grail as the object of the Knightly Quest is the most central and mysterious theme. In these stories, the Grail is the key to restoring goodness to the land. Despite the Grail's mythical reputation, many scholars believe that it may be something quite real. The Grail romances most commonly depict the Grail as the cup of the Last Supper of Christ into which the blood of Jesus dripped from the cross. In turn, the Grail generally is supposed to possess magical powers including the ability to cure all wounds and bestow immortality upon those deemed worthy of its divine gifts. The Grail is described as a source from which an endless wealth of all manner of desirable food may come. Even more significantly, all life on earth proceeds from the Grail including all produce from the ground, fish, fowl, and animals. Whether the Grail is the actual cup of Christ or a symbol for something else entirely has been a topic of debate for centuries. In the 12th and 13th centuries, the Grail mystery reached such popularity that it was virtually a sub-cult of Christianity in England, France, and Germany.

CHAPTER 2

Flegetanis the Astronomer

He declared there was a thing called the Grail,
whose name he had read clearly in the constellations.
Parzival

The most significant of all the Grail romances is "Parzival" composed between 1195 and 1216 by a Bavarian knight, Wolfram von Eschenbach. Nowhere in "Parzival" is the Grail ever associated with a cup. In this tale, the Grail is uniquely and fascinatingly a stone. Wolfram claims he based his version upon privileged information he received from a scholar from Provence, France named Kyot. Kyot himself received his knowledge from an astronomer. This privileged knowledge is unique to Parzival, and indicates the central thesis of a secret scientific mystery. The source of this knowledge is a manuscript found by Kyot. The manuscript was laying neglected in the corner of a library in Muslim possessed Toledo, Spain. The author of the text was

Flegetanis, an astronomer who wrote of the marvels of the Grail.

The first clear point to make about this medieval text is that the astronomer Flegetanis is Wolfram's primary source of Grail knowledge. All rests upon him for whatever plausibility Wolfram von Eschenbach may have in delivering the Grail secret to the world in carefully encoded symbolism. Wolfram claims knowledge of the secret of the Grail from Kyot of Provence, and Kyot gets the secret of the Grail from the book written by Flegetanis. The book was written in a "heathen" language, and Kyot had to learn this foreign language in order to read it. Wolfram tells us that Flegetanis had a Hebrew mother and a heathen father and lived in Muslim Spain. As Flegetanis is emphatically not a Christian, it serves to reason that the marvel of the Grail might not actually seem to be specifically Christian.

What exactly was it about the manuscript that so inspired Kyot of Provence to learn a new language in order to read it? Since at the start, Kyot could not read the book, he was most likely interested in the illustrations. Considering that Flegetanis was an astronomer, Kyot may have recognized forbidden astronomical knowledge. These could have been diagrams as simple as depicting the Sun, our local star, encircled by the planets. It is now obvious to everyone that the planets orbit around the sun, yet scientists were burned at the stake for promulgating such ideas. This was the fate of men like Giordano Bruno (1548-1600). He was gagged and hung upside down naked in the Campo de' Fiori in Rome and burned (in part) for his promotion of the Copernican view, putting the sun at the center of the solar system. The following passage sets Parzival apart from all the Grail romances because it is explicit in providing concrete clues leading to the Grail.

> For the infidel Flegetanis was able to define for us the recession of each planet and its return, and how long each revolves in its orbit before it stands at its mark again. All human kind are affected by the revolutions of the planets. *With his own eyes* the heathen Flegetanis saw - and he spoke of it reverentially - hidden secrets in the constellations. He declared there was a thing called the Grail, *whose name he had read clearly in the constellations.* A host of Angels left it on Earth.[4]

The text says Flegetanis saw "with his own eyes," hidden secrets in the constellations, and that a host of angels brought the Grail, "whose name he had read clearly in the constellations," from heaven to earth. If these things were images that were forbidden knowledge within Christian culture, then they may have instigated Kyot of Provence to learn the language in order to be able to read the manuscript. No one has ever before published a Grail theory based upon this passage about Flegetanis or identified these "hidden secrets" and the "name" of the Grail. *Wolfram's passage sets his work apart and reveals the secret that will make it possible for a determined individual, like Kyot of Provence, to find the Holy Grail.* The first step would be obtaining sufficient astronomical knowledge of the planets to have the ability to mark or trace planetary orbits around the sun. Thus, the determined individual has a great task to test their mettle. This assignment would be daunting to any noble knight.

Of all the authors that have written on the topic of the Grail, Arthur Edward Waite is the only one who comes close to this thesis of the Grail by recognizing the importance of Flegetanis independently of the entire corpus of Grail literature. About Flegetanis he writes.

---

[4] Wolfram Von Eschenbach, A. T. Hatto (Translator), *Parzival.* (London, England: Penguin Books Ltd, 1980) p.232 (Author's italics).

> I rule therefore that much needs to be said for the clear sight of that Son of Israel and Paganism who found the Graal – record in a galaxy of stars, and though the method by which that record is decoded will not be found in the course of a day's reading at any observatory, I am quite sure that the stars will still tell the same story, that it is also the true story, which owes nothing to the Chronicles of Anjou.[5]

By writing that the way of decoding the secret of the Grail may not be found in a day, Waite acknowledges the difficulty inherent in deciphering exactly what it was that Flegetanis saw in the constellations, and he intuits that the stars still show the exact same phenomena today, meaning that it repeats throughout history, unlike perhaps, a supernova occurring on a specific date. This means that what Flegetanis saw, with his own eyes and in his own day, was not a chance historical occurrence such as a one time or rare planetary and celestial alignment (such as some theorize the Star of Bethlehem to be) occurring in the distant past. Some twenty-four years later, A.E. Waite also wrote, "the story of Flegetanis has suffered what is contamination by the introduction of extraneous matter, being all that which was not included in the record of the starry heavens."[6] In other words A.E. Waite intuited, without knowing what it was, that the key to the Grail secret is astronomy and all else is superficial! He wrote that, "When the Jew of Toledo read in the great sky, as in a glass of vision, it does not mean that he arranged the fixed lights into conventional forms, but that he divined as a devout astronomer."[7] By this he is referring to the zodiacal diagrams by which human imagination has

---

[5] A.E. Waite, *The Hidden Church of the Holy Graal – It's Legends and Symbolism Considered in Their Affinity With Certain Mysteries of Initiation and Other Traces of a Secret Tradition in Christian Times.* (London: Rebman Limited, 1909). p.396.

[6] A.E. Waite, *The Holy Grail – It's Legends and Symbolism.* (London: Rider and Co. 1933, Mineola, New York: Dover Publications Inc. 2006). p.272.

[7] Waite, *The Hidden Church.* p.396.

assigned pictograms to star patterns in the sky. But when Waite writes that Flegetanis did not use "conventional forms" he is meaning that the wonders Flegetanis saw were nothing so imaginative or mythical like the familiar and ancient zodiacal symbols. He also wrote, "The mystery which the stars expressed is that by which, in the last consideration, all the material planets are themselves ruled."[8] Meaning that the mystery is not astrology but true empirical science!

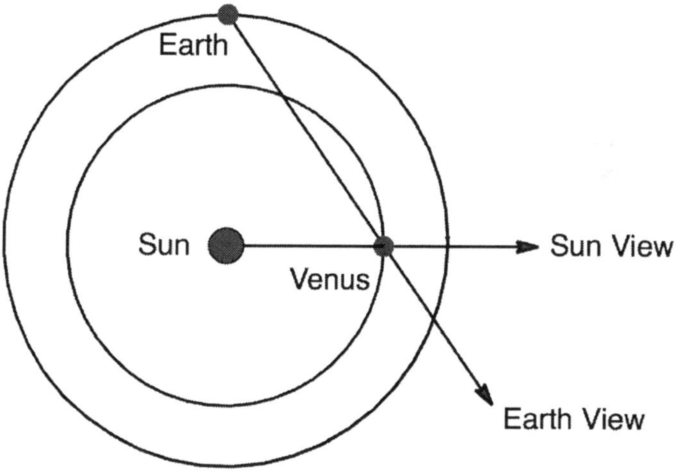

Figure 1 The Error of Geocentric Perspective

Might it be that the Catholic Church made a sun-centered cosmology a heresy in order to prevent anyone from seeing these hidden secrets of Flegetanis? Consider the literal perspective of looking at the constellations from the surface of the Earth or the alternative viewpoint of the sun. The problem in perspective is the Earth you stand on is constantly both moving in orbit around the sun and rotating at the same time. Imagine that you look at Venus

---

[8] Ibid. p.396.

in the evening sky and see that it appears in a well-known zodiac star pattern such as “Pisces,” the two fish symbolic of the Piscean Age in which the precession of the equinox has been for approximately 2,000 years. The earth-centered viewpoint would be Venus in Pisces. It is not easy or obvious to see the position of Venus as you would see it if you were standing upon the surface of the Sun. As shown in the diagram, this would be a very different area of the sky and the background may be a different star pattern such as “Virgo.”

The first modern astronomer to fully understand the motion of the planets was Johannes Kepler. Throughout his voluminous writings, he stated repeatedly that the correct perspective for viewing the planets was as though your feet were standing on the surface of the Sun. In order to change his viewpoint of the heavens, so that planetary positions made sense, he used geometry and trigonometry. Therefore, if Flegetanis gave instructions on reading the “name of the Grail in the constellations” then the constellations would be different given the perspective used with the Sun at the center. An entirely different view of the heavens will be the result.

The theory of the Grail explained here provides the only concrete reason as to why the medieval Catholic Church made the concept of a heliocentric solar system a heresy. By forcing where society must stand on this one issue of astronomy, they prevented the key to the secret of the Grail from advancing into the world, and it could only lay in repose. There are enormous implications if the secret of the Grail may be true astronomical knowledge. If the Roman Empire had stood upon this perspective, then the world may have avoided the intervening Dark Ages and proceeded directly into the Age of Enlightenment. The societal foundation stone must be established upon science and reason and not the shifting sands of beliefs. As it is written, it is only “a foolish man who built his house upon sand”. (Matt 7:26)

CHAPTER 3

Neutral Angels

I know your deeds,
that you are neither cold nor hot;
I wish that you were cold or hot.
So because you are lukewarm,
and neither warm nor cold,
I will spit you out of my mouth.
Rev 3:15-16 NASU

When Wolfram von Eschenbach first presents the Grail in his story, he gives no detailed description of its actual appearance, which heightens its mystery. The Princess of the Grail family, named Repanse de Schoye, wearing a dress of Arabian silk, bears the Grail on a Green achmardi (perhaps a cushion). The Grail is described as, "the consummation of heart's desire, its root and its blossoming [...] transcending all earthly perfection." The theme developed by Wolfram von Eschenbach, is that during the mythic war of the Angelic Rebellion a host of angels brought the Grail to Earth from heaven and protected it until it became the responsibility of worthy humans. According to Wolfram:

> When Lucifer and the Trinity began to war with each other, those who did not take sides, worthy, noble angels, had to descend to Earth with that stone which is forever incorruptible. I do not know whether God forgave them or damned them in the end: if it was His due He took them back.[9]

Neutral angels brought the Grail, which is called an incorruptible stone, to Earth to keep it safe during the time of war. When war rages there is little place for those who choose to remain neutral. Wolfram questions whether God damns the angels or forgives them for their neutrality. At the conclusion of his epic story, he refers back to this passage and concludes that they cannot be forgiven because God is constant in His animosity. It seems the Grail was a casualty of an ideological war. Christianity propagated itself while simultaneously destroying other myths and religions before it. However, it is clear the Grail was something that was too valuable for those with more *level heads* to really destroy.[10]

The idea that neutral angels preserve the Grail gives us some clue as to its nature. Perhaps it is true knowledge made forbidden because it does not polarize people. It could be a uniting, undeniable truth, made more potent by its mystical and divine implications. The indication is that the Grail was something known in the distant past that was deemed an impediment to the ideological propagation of Christianity. The neutral angels, whom Wolfram calls worthy and noble, preserved this knowledge. These "neutral angels" are symbolic of very real people in the ancient past. It is clear they were people with open minds who protected the secret of the Grail by remaking it mythically into something suitable for Christianity. The

[9] Eschenbach, *Parzival*. p.240

[10] Being "on the level" is a phrase originating in Freemasonry, and it means fairness and honesty.

following is Wolfram's description of the Grail as a Stone, and of those who defended it in the Grail Castle.

> Well, I know that many brave knights dwell with the Grail at Munsalvaesche. Always when they ride out, as they often do, it is to seek adventure. They do so for their sins, these Templars, whether their reward be defeat or victory. A valiant host lives there, and I will tell you how they are sustained. They live from a stone of purest kind. If you do not know it, it shall be here named to you. It is called "Lapsit exillis." By the power of that stone, the phoenix burns to ashes, but the ashes give him life again. Thus, does the phoenix molt and change its plumage, which afterward is bright and shining, and as lovely as before. There never was a human so ill but that, if he one day sees that stone, he cannot die within the week that follows. And in looks he will not fade. His appearance will stay the same, be it maid or man, as on the day he saw the stone, the same as when the best years of his life began, and though he should see the stone for two hundred years, it will never change, save that his hair might perhaps turn gray. Such power does the stone give a man that flesh and bones are at once made young again. The stone is called the Grail.[11]

The Grail knights are sustained from this stone called the Grail. If any should see the Grail, whether ill or mortally wounded, they will not die for at least a week. If seen in normal health, the Grail may sustain them for two hundred years, with only their hair turning gray. The power of this stone specifically enables the phoenix to burn to ashes and be reborn again to full plumage. The phoenix that rises from the ashes is a symbol of esoteric wisdom that cannot be destroyed despite the recurring historical destruction of knowledge. The Phoenix mentioned in the quote above is closely associated mythically to the Philosopher's Stone, which might be interpreted as the periodic restoration and rebirth of

[11] Ibid p.239

esoteric wisdom in the world. The point here specifically about the Grail stone is that it plays a specific role in both the destruction and resurrection of esoteric knowledge. The fact that the Phoenix will rise from its ashes and appear as lovely as it did before it was burned indicates the secret of the Grail. Although it is a truth that was inconvenient in early Christian times, it may be restored in the current day to its former glory.

Despite the centuries of condemnation by God or an earthly Church, Wolfram makes it clear that the Grail has a higher divine favor bestowed on it. On every Good Friday, a dove depicting the Holy Spirit, flies down from heaven bringing a small white wafer to the stone and then returns to heaven. It is this wafer that gives the stone its magical powers. Wolfram narrates:

> Every Good Friday, as I say, the Dove brings it to the stone, from which the Stone receives all that is good on Earth of food and drink, of paradisal excellence – I mean whatever the Earth yields. The Stone, furthermore, has to give them the flesh of all the wild things that live below the aether, whether they fly, run or swim - such prebend does the Grail, thanks to its indwelling powers, bestow on the chivalric brotherhood.[12]

The word "prebend" is the same in meaning as a stipend or dividend and infers a periodic payment. Good Friday occurs once every year and the Holy Spirit "invests" something in the Grail, for which in turn it pays out to its guardian Templar Knights food and drink of Paradisal excellence. The symbolism of the Phoenix has a parallel in Christianity. Since Good Friday is the day of Christ's Crucifixion, then Christ is like the phoenix that burns to ashes, but which gains life from the ashes, and is resurrected in three days. Some have suggested that Christ

---

[12] Ibid, p.240

is the stone that is the Grail. The dove is clearly a symbol of the Holy Spirit, but what is the symbolic meaning of the wafer? What does the dove bring to the stone that makes the stone produce all living things on Earth? The Grail functions similarly to the ancient Greek mythical concept of a cornucopia or horn of plenty. This symbol of abundance dates back to the fifth century B.C.E. The cornucopia was the mythical horn of the nymph Amalthea containing food and drink in endless supply. It is a non-complex interpretation of the Grail to hint at something of cosmic significance. Because of its communication with the Holy Spirit and receiving the wafer, the Grail stone produces "whatever the Earth yields," meaning all life, including birds, mammals, and the fish in the seas. This *Holy Communion* is a mystical meeting point between science and religion. There is something in common between life on Earth and the heavens above and the abstract symbol of this truth is the Grail.

What we call the Grail long predates the time of Jesus. There is evidence that humans have known about it for at least five thousand years, making it a sort of very ancient Holy Relic. It is mathematical and scientific knowledge that provides an insight into the genesis of life on earth and its correspondence with cosmological harmonies in the solar system. Wolfram writes that people who are not pure of heart cannot tolerate being in the presence of the Grail. To perceive it correctly you must also be open minded and prepared to dispense with some long held prejudices that you may have been taught since childhood. There are a couple of very uncomfortable facts about the Grail. It will become evident that the Grail secret has long been associated with witchcraft and the antichrist. This degree of prejudice is why Wolfram von Eschenbach declared at the conclusion of Parzival that the neutral angels would never be forgiven because, "God is constant in such matters." Therefore, only those with the eyes to see and the ears to hear, may understand the Holy Grail as deserving of divine grace.

CHAPTER 4

A Stone From Lucifer's Crown

He took Lucifer's crown from his head,
in such a way that a stone jumped out of it,
which on earth became Parsifal's stone.
Wartburg War (1207 AD)

In all mythologies, and especially theologies, truth is hidden behind a veil and symbolism may be the inverse image of truth, as though seen in a mirror, reversing the left and right perspective.[13] To properly interpret symbolism you may need to be able to see it in a way that is different from what is commonly taught. Since in Wolfram von Eschenbach's Parzival the Grail is a stone or rock, we need to consider what is unique about this symbolism. For example in the time of Jesus, there were no temples or great cathedrals for the Christian faith except for those existing in the imagination of people. In the New Testament, believers are encouraged, "like living

[13] Related mirror symbolism is used in the separate myths of the Medusa and vampires.

stones be yourselves built into a spiritual house" (1 Peter 2:5).[14] The kingdom that is not of this world has a temple that is also not of this world. The believers are the spiritual stones of this temple. If the spirit is like a rock, then the soul is likened to that which is most solid and indestructible. However, elsewhere in scripture a rock causes some to stumble and is a trap for those who do not suspect it. "And whoever falls on this stone will be broken; but on whomever it falls, it will grind him to powder." (Matt 21:44 NKJV)[15] Dogmatists interpret this stone as Christ. The conservative Bible commentator Matthew Henry (1662-1714) writes, "Those who reject him as a Savior will split upon him as a Rock. [...] All those who go on resolutely in their infidelity and contempt of the gospel are appointed to eternal destruction; and God from eternity knows who they are."[16] Christian dogmatists construe those who stumble as unbelievers because the common interpretation of the "stumbling stone," is to apply this to the concepts of damnation and salvation. The ultimate goal of Christian salvation is mass cultural assimilation. Matthew Henry writes, "All Christians, wheresoever they be, compose one holy nation [...] agreeing in the same manners and customs, and governed by the same laws."[17] Theologians try to trip the unwary into their "briar patch" of complex psychological traps and snares, and the stone that will fall on unbelievers and grind them to powder is assumed as a metaphor of this. However, the Grail stone is associated with Lucifer and the rebellious angels and represents something entirely in opposition to submission to authority. In the Grail epic,

[14] See also Mark 12:10, Matt 21:42, Acts 4:11, 1 Peter 2:4-9, Isa 8:13-15.

[15] See also Isaiah 8:15.

[16] Matthew Henry, *Matthew Henry's Commentary on the Whole Bible Complete and Unabridged in One Volume* (Peabody, Massachusetts: Hendrickson Publishers Marketing LLC, 1991, 2008). 1 Peter 2:4-12, p.1948.

[17] Ibid. 1 Peter 2:4-12, p.1948.

"Parzival" by Wolfram von Eschenbach it is made clear that the Grail is celestial in origin. For Wolfram the Grail is a stone called "Lapsit exillis," or in some translations, "lapis exilis." This name, which resembles Latin, is not. Latin scholars have suggested that he derived the name from a Latin phrase like "lapsit ex caelis" - it fell from the heavens.

In 1207 Wolfram von Eschenbach was in the court of Hurmann, Landgrave of Thuringia at Wartburg Castle near Eisenach in central Germany. This is the same castle that Martin Luther, the famous Protestant and founder of the Lutheran Church was given asylum. In 1521 Martin Luther's life was put in danger after being excommunicated by a Papal representative at the Diet of Worms. He was excommunicated for challenging the authority of the Roman Church and at Wartburg Castle translated the New Testament into German. As Martin Luther was given asylum at the same Wartburg Castle where Wolfram was three centuries before, it betrays a long-term association of Wartburg Castle as a stronghold for heretical challengers of the Christian faith. Further research in this regard may prove rewarding. In 1207 Wolfram was there engaged in a poetic epic battle with another famous poet, Heinrich von Waldeck. In the resulting epic poem "Wartburg War" strophe 143 we find mention of the Grail with the tale of the Angelic Rebellion. These are the words of Wolfram von Eschenbach, which when combined with the symbolism in Parzival, are a key to understanding his secret of the Grail.

Shall I then bring the crown,
that was made by 60,000 angels?
Who wished to force God out of the Kingdom of Heaven.
See! Lucifer, there he is!
If there are still master-priests,
Then you know well that I am singing the truth.
Saint Michael saw God's anger, plagued by this insolence.

He took Lucifer's crown from his head,
In such a way that a stone jumped out of it,
Which on earth became Parsifal's stone.
The stone which sprang out of it, he found it,
he who has struggled for honor at such high cost.[18]

Most people in the world believe religious myths are immutable and eternal, but the fact is they change over time. Lucifer is wrongly associated with Satan, but most people are taught Lucifer is evil and would be uncomfortable with any other idea. The following similar account appears in "The Secret Teachings of All Ages" by Manly P. Hall, the founder of the Philosophical Research Society.

> Like the sapphire Schethiya, the Lapis Exilis, crown jewel of the Archangel Lucifer, fell from heaven. Michael, archangel of the sun and the Hidden God of Israel, at the head of the angelic hosts swooped down upon Lucifer, and his legions of rebellious spirits. During the conflict, Michael with his flaming sword struck the flashing Lapis Exilis from the coronet [crown] of his adversary, and the green stone fell through all the celestial rings into the dark and immeasurable Abyss. Out of Lucifer's radiant gem was fashioned the Sangreal, or Holy Grail, from which Christ is said to have drunk at the Last Supper.[19]

Lucifer has associations that many people are unfamiliar with. For instance, the name Lucifer means light bringer (luce light + ferre bring). There is a cave in the south of France in Saint-Baume that in Roman times was the center for worship of the goddess Diana Lucifera the light bringer or Illuminatrix. Lucifer is also the Latin name of the second planet from the sun; more commonly

---

[18] Karl Simrock, *Der Wartburgkrieg.* (Cotta, Stuttgart, Augsburg 1858. OCLC 5363166), a German translation of the medieval *Codex Manesse, Janaer Leiderhandschrift* and translated into English.

[19] Manly P. Hall, *The Secret Teachings of All Ages – An Encyclopedic Outline of Masonic, Hermetic, Qabbalistic, and Rosicrucian Symbolical Philosophy.* (Los Angeles, California: Philosophical Research Society, Inc. 1929, 1988). p.XCIX.

called Venus after the goddess of love. The mythic Lucifer that most people think of as a malevolent male supernatural being therefore has mythic associations with goddess worship. The conflict of Christianity with earlier goddess worship is likely to be part of the ideological conflict that the myth of Lucifer and the Angelic Rebellion represents. Lucifer is said to have been the most beautiful of angels, and in the angelic rebellion, the angels that joined him/her (Lucifer/Venus) may just be symbols for the people who attempted to preserve the mythology and science of the ancient world from destruction during the rise of Christianity.

Consider this Old Testament verse, "The stone which the builders rejected is become the head stone of the corner. This is the LORD's doing; it is marvelous in our eyes." (Ps 118:22-23 KJV) In common Christian scriptural interpretations, the stone is Christ and the builders are identified with rulers of the Jews who reject Christ. However, the "builders" are the builders of a spiritual temple. With the aid of unknown agents somehow the stone is promoted despite their rejection. In Wolfram's Parzival, these would likely be the "neutral angels." In this Grail thesis, *the stone rejected by the "builders" is not Christ, but the Grail secret rejected to this day by the Christian theologians themselves.* Theologians will reject truths found in nature that seem to contradict their mythology. The dogmatists and theologians are the very people who become blinded to this obvious interpretation, as they are primarily accustomed to pointing their fingers at others. The theologians are not able to see themselves in this mirror precisely because they are self-righteous, which is the delusion of all those who are dogmatic and trained to not question what they are told. Also, in this Grail interpretation, Lucifer seen in the mirror of alternative interpretation will transform from a malevolent fallen male angel into the goddess of love and wisdom (Sophia). As the apple supposedly did to Isaac Newton; perhaps the Holy Grail as a stone falling from heaven, could make quite an impact upon the heads of dogmatic theologians?

CHAPTER 5

The Secret of God's Power on Earth

I will give you the keys of the kingdom of heaven,
and whatever you bind on earth shall be bound in heaven,
and whatever you loose on earth shall be loosed in heaven.
Matthew 16:19 RSV

It has been said the Holy Grail guards the secret of God's power on Earth. This is quite true and it is an obvious reason for it being classified as heretical by the theologians who wield God's power. Most heretical ideas can be understood as undermining the power of theocracy over the masses of people. God's power on Earth derives from a deceptively simple idea. You have almost certainly heard it expressed in one form or another. In its most pure and ancient form it is known as the Hermetic Principle of Correspondence, "As above, so below; as below, so above." The Hermetic Principle of Correspondence is the key to understanding how all theology and dogma work their magic. This comes from the Hermetic wisdom named after the legendary sage

Hermes Trismegistus.[20] If you are not familiar with this Hermetic Principle you may recognize it in the Lord's Prayer, (Matt 6:9-13) which contains the words, "Our Father which art in heaven, hallowed be thy name. Thy kingdom come. *Thy will be done on earth as it is in heaven.*" The relation to the Hermetic Principle should be obvious. The blueprint for how the world is envisioned to be, is conceived in the concept of heaven above. For example calling God in heaven our "Father" is a mythic idea that anthropomorphizes God. It changes the form of the divine into human (anthropos) form.[21] It makes obvious the power of the Hermetic principle in its practical application. By promoting obedience to God and making God in heaven our "Father," we put human men in positions of power and authority on earth below. Prayers such as the Lord's Prayer convince the masses to accept and obey authority from men.

Theocratic power hinges on controlling what we, in our imagination, perceive in heaven above. It is the Archimedean lever moving the great herds of humanity on earth. This is why the medieval theologians never took kindly to anyone besides themselves claiming any authority about what is in heaven above. The heavens above were the jurisdiction of the clerics and in the Middle Ages, this primacy was protected with fire and sword. This was the cause for the persecutions of scientists and especially astronomers dealing in heavenly matters. Therefore, if the Grail secret is an inconvenient but objective scientific truth displayed in the heavens above, then all who have the correct perspective ("those with the eyes to see") may see it and know that religion, as a

---

[20] Three Initiates. *The Kybalion, A Study of The Hermetic Philosophy of Ancient Egypt and Greece.* (Masonic Temple Chicago, Ill.: The Yogi Publication Society, 1912) p.113.

[21] Greek *anthropomorphize* is "anthropos" = human + "morphe" = change to human form.

product of man, is but an imperfect reflection of the divine.

The Roman Catholic Church has claimed its authority from apostolic succession of the bishops of Rome through Peter. Church authorities have used Matt 16:18 for centuries as the foundational stone, or rock, upon which they built this claim. It was suggested earlier, the verse Matthew 16:18 is a symbolic key to understanding the solution to the Grail secret. It is now provided in full context for your consideration.

> 13 Now when Jesus came into the district of Caesare'a Philip'pi, he asked his disciples, "Who do men say that the Son of man is?" 14 And they said, "Some say John the Baptist, others say Eli'jah, and others Jeremiah or one of the prophets." 15 He said to them, "But who do you say that I am?" 16 Simon Peter replied, "You are the Christ, the Son of the living God." 17 And Jesus answered him, "Blessed are you, Simon Bar-Jona! For flesh and blood has not revealed this to you, but my Father who is in heaven. 18 And I tell you, you are Peter, and on this rock I will build my church, and the powers of death shall not prevail against it. 19 I will give you the keys of the kingdom of heaven, and whatever you bind on earth shall be bound in heaven, and whatever you loose on earth shall be loosed in heaven." 20 Then he strictly charged the disciples to tell no one that he was the Christ. Matthew 16:13-20: RSV

In The Catholic Encyclopedia we read about Matt 16:18 and about Peter that, “He should be, next to Christ himself, the chief foundation stone, in quality of chief pastor, ruler, and governor; and should have accordingly all fullness of ecclesiastical power, signified by the keys of the kingdom of heaven.”[22] Note the significance of the “keys of the kingdom” from verse nineteen that denotes power. The Protestant commentaries are very careful in how they deal with Matt 16:18-19 because it is of such key importance to

[22] www.newadvent.org

priestly power. The Roman Catholic Church essentially "owns" Saint Peter because Saint Peter's Basilica in the Vatican is literally built upon Peter's tomb. For instance in Adam Clarke's Commentary, "Jesus Christ did not say, on thee, Peter, will I build my church, but changes immediately the expression, and says, upon that very rock, [...] to show that he neither addressed Peter, nor any other of the apostles. So, the supremacy of Peter, and the infallibility of the church of Rome, must be sought in some other scripture, for they certainly are not to be found in this."[23] Some may wonder why Peter would be the rock upon which the Church would be built. He is the one Jesus predicted would deny him three times before the cockcrow on the morning following the last supper (John 13:38, Luke 22:34). Also just following the verse above (Matt 16:23), after Peter tries to tell Jesus that he does not need to die on the cross Jesus says to Peter, "Get behind me Satan. You are a hindrance to me." The most common Protestant interpretation is that the rock is Christ himself. Others interpret that what Peter confesses, "You are the Christ, the Son of the living God," is the rock. This is a core foundational idea of the Christian religion. Matthew Henry says that without this truth, the universal church will fall to the ground because Christianity would then be a cheat and the church a mere "chimera," a mythical monster. He writes, "This is *articulus stantis et cadentis ecclesia—that article, with the admission or the denial of which the church either rises or falls;* 'the main hinge on which the door of salvation turns.'"[24] Matthew Henry identifies the Catholic Church as the builders of the church that "trip and fall" and reject Christ. Perhaps regarding the evils perpetrated by the Catholic Church in

---

[23] *Adam Clarke's Commentary, Electronic Database.* Copyright © 1996, 2003, 2005, 2006 by Biblesoft, Inc. All rights reserved. (On Matt 16:18).

[24] *Matthew Henry's Commentary* Matthew 16:13-20, p.1348

the medieval period he writes, "Still the hierarchy of Rome is the worst enemy in the world to Jesus Christ and his interest."[25] Perhaps you get a sense of the sacred ground upon which we tread, and how dangerous it may seem to offer up a new dimension of Matt 16:18 for both Catholics and Protestants. The "keys of the kingdom of heaven" (Matt 16:19, Matt 18:18) are normally interpreted as the keys of power being the establishment of the institutions of the Church. The Protestant Matthew Henry writes, "It is not any civil, secular power that is hereby conveyed, *Christ's kingdom is not of this world;* [...] It is a power to *bind and loose,* that is (following the metaphor of the keys), to shut and open."[26] He identifies the keys as doctrine and the minister's power to admit, or expel people from their church membership. Note that the verse of Matt 16:19 uses the plural "keys" and there are seven Hermetic principles which may be the very same keys mentioned. The source of theocratic power rests upon mythology, and this explains the fierce rejection of any erosion in interpretations of religious dogma. Theological debate is unlike other forms of debate in that it is subjective and symbolic and with strict rules of engagement. The ministers wield a power like a form of mental Jiujitso (a gentle art of combat without weapons), with the power to bind or loose. Theology is, in many ways, a form of martial artistry with words.

The Catholic Church has promoted the dogma of apostolic succession from Peter the disciple and this book's interpretation of the Grail obviates the importance of apostolic succession by providing a never before published interpretation of "Peter the rock" as a key to the secret of the Grail, and this is central to the argument of the Grail being the rejected stone and heretical.

---

[25] Ibid. 1 Peter 2:4-12, p.1948.

[26] Ibid. Matthew 16:13-20, p.1348.

CHAPTER 6

Repanse de Schoye

King Arthur beheld all the changes;
the last whereof was the change into a chalice.
Perlesvaus

Manly P. Hall wrote that there has never been an adequate interpretation of the Grail mysteries. However, he believed the Knights of the Holy Grail were a powerful organization of Christian mystics perpetuating ancient wisdom, and that the Grail quest is the eternal search for Truth. He considered King Arthur to be the Grand Master of a Christian-Masonic brotherhood and that Arthur was in his position because he accomplished the withdrawal of the sword (spirit) from the anvil of base metals (his lower nature). In his opinion, the historical Arthur became confused with the allegories and myths of his esoteric mystical order and he identified the Grail with the Philosopher's Stone. According to Manly P. Hall in "The Secret Teachings of All Ages."

> The philosopher stone is an ancient symbol of the perfected and regenerated man [...] In philosophy the Stone of the Wise Man [according to Albert Mackey] is supreme and unalterable Reason. [...] He who possesses the Philosopher's Stone possesses Truth, the greatest of all treasures, and is therefore rich beyond the calculation of man; he is immortal because Reason takes no account of death, and he is healed of Ignorance - the most loathsome of all diseases.[27]

Chretien de Troyes wrote the earliest written work connecting the Grail with King Arthur and his knights. This was "Percival or The Story of the Grail," written by him in the late 12th century. Chretien was a French cleric and poet born about 1135 and he lived during the time the Crusaders still held Jerusalem. In his story, the main character Percival was something of a lout. He leaves his widowed mother and sallies forth to win his knighthood. During his journeys, he comes upon an enigmatic fisherman in a boat on a lake. He is the wounded "Fisher King," who offers him refuge for the night in his castle. The wound of the Fisher King does not allow him to sit upright or stand; he may only recline to not incur pain. During the evening, a procession of maidens and squires precede a feast. A squire carries a white lance by the middle of its shaft and from the tip of the lance, a large drop of blood issues forth and flows down to the squire's hand. Then two squires carry golden candelabrum with ten candles each. A maiden carries in the Grail, holding it with both hands, then another maiden behind her, carries a silver carving-platter. In this version, the Grail is described as being made of gold and inlaid with many fine precious gems. However, Chretien does not say what it actually is, nor does he link the Grail in any specific way to Jesus. This version of the Grail story is similar to all the

---

[27] Hall, *Secret Teachings*. p.XCVI

others in the following way. Percival is supposed to ask regarding the Grail, "Whom does it serve?" Out of politeness and respect for the Fisher King, he fails to make this bold query on his first encounter with the Grail. However, the asking of the question itself is the key to attain the Grail secret and heal the wound of the Fisher King. In all versions, Percival always has to pass many trials before getting another chance to be in the presence of the Grail and the Fisher King. *One obvious interpretation of this element of the Grail stories is that special knowledge only goes to those who know what questions to ask.*

Chretien's version of the Grail story ends abruptly in mid-sentence, which possibly coincides with the author's death in 1188. This was shortly after the Crusader's loss of Jerusalem on July 4, 1187 when Saladin defeated Guy of Lusignan, then King of Jerusalem. Some authors propose that Chretien may have been killed because he was writing about the Grail, but if this were true, one has to wonder why the book itself did not vanish. The Chretien de Troyes version of the Grail lore served to fire the collective imagination of the 13th century and the Grail became deeply woven into western culture, as it is to this day. Despite Catholic clerical condemnation, the Grail romances flourished for about a century. Then in the fourteenth century, beginning about 1307 and coinciding with the persecution of the Knights Templar, the Grail stories largely vanished from the stage of European history for nearly two centuries. It was at this time as well, that the Knight card was removed from the playing cards, which to this day only have the King, Queen, and Squire (who served the Knight). The Englishman Sir Thomas Mallory in his "Le Morte D'Arthur" or "The Death of Arthur" reworked the Grail theme about three centuries after Chretien de Troyes. This much Christianized version of Arthurian legend was published in 1485 and has ever since been a part of the tapestry of Western culture. How coincidental might it be, that the "Malleus Maleficarum" or "Witches Hammer" was first published in 1487, only two years after the publication of this Grail romance? This most infamous of books was the prescribed text for the witch persecutions of the following era. It is possible that the resurfacing of the Grail at this time in history instigated the "Burning Times," or the persecution of witches in response.

The Grail changed form over time. For instance, a French monk named Helinandus wrote in his chronicle of 1204 that the Grail was the bowl that the Lord had used at the Last Supper with his disciples. It was also described as a platter, kind of like a cornucopia from which an endless supply of food and drink may come. By Sir Thomas Mallory's time, it had somewhat rigidly assumed the distinct identity it still has today. It is supposed to be the cup of the Last Supper in which Joseph of Arimathea later caught the blood of Jesus during the crucifixion. According to some accounts, Joseph of Arimathea brought the Grail to Glastonbury England. By other accounts, it was brought by the Magdalene to southern France.

Another telling of the Grail written between 1190 and 1212 makes it very clear the Grail involves a big secret in some way related to Jesus. This is the "Perlesvaus" written by an anonymous author. It was unusual then to seek anonymity, since poets of the day enjoyed an exalted status. Therefore, others have theorized that the author may have been in a position for which it was not appropriate to write about the Grail, such that this author may have been a member of a monastic or military order. It is possible the author was associated with the Catholic Church, which disapproved of the Grail story. Its French author claimed he based the "Perlesvaus" upon a Latin book written by a monk of Glastonbury, significantly, where the Grail is supposed to have been delivered by Joseph of Arimathea. Since in the story the knights of the Holy Grail wear mantles embroidered with red crosses like the mantles the Knights Templar wore, and since the author demonstrates detailed knowledge of war, weapons, armor, and military strategy - some think the author may have been a Templar Knight. Some scenes may be interpreted as having Masonic symbolism, for example, thirty-three men, dressed in white with red crosses on their chest, who all appear to be thirty-two years of age. In "Morals and Dogma" by Albert Pike there are thirty-two degrees and "33" is in the double-headed eagle Masonic symbol and possibly describes a symbolic degree.

In the Perlesvaus, the Grail is a secret of Christ. Gawain, a knight on the Grail Quest in this story, is warned by a priest, "[...] for no-one should reveal the secrets of the Savior; they should be kept secret by him to whom they are entrusted."[28] In this version, the Grail is a changing sequence of images. Gawain perceives different images within the Grail - a chalice, a child, and a crowned king crucified. His attention is raptured and he fails to ask the question that will heal the Fisher King.

Later in the book, King Arthur, Gawain, and Perlesvaus return to the castle of the Fisher King who is deceased. There is a stream that passes behind the castle which comes from Earthly Paradise, which is filled with fish. Wherever, "the stream flowed there was a great abundance of all good things."[29] The stream then flows on into the forest to the home of a worthy hermit where it vanishes into the earth. This concept of an underground stream is commonly symbolic of nourishing esoteric

---

[28] Nigel Bryant, *The High Book of the Grail – A translation of the thirteenth-century romance Perlesvaus* (Cambridge, D.S.Brewer, Revised Edition 2007) p.63

[29] Ibid p.195

traditions that exist in the cultural underground. For example in the "Aeneid" by Virgil (Book III, 694-696), the Arcadian Stream is said to have run under ground and also under the sea to resurface at Ortygia in Sicily. In the "Perlesvaus," the Grail is multiple things and changes form. King Arthur attends a mass in which:

> The Grail appeared at the consecration in five forms, but they should not be revealed, for the secrets of the sacrament none should tell save the one to whom God had granted grace. But King Arthur saw it in each of its forms in turn, and last appeared as the chalice;[30]

Most Grail theories do not explain this multi-faceted changing nature of the Grail. Neither do other Grail theories explain how it is that the astronomer Flegetanis saw, "and he spoke of it reverentially - hidden secrets [*note the plurality*] in the constellations."[31] Any scholar who studies the Grail must come to ask the question, is it possible there is a lot more to the Grail than has been previously thought about? Could there be some world-changing secret, of something very real, that is the Grail? The changing symbolism in the "Perlesvaus" may suggest that those privileged people of that age who knew the secret of the Grail, sought to preserve it from persecution by encoding it in Christian symbolism. As though a committee were debating the merit of different symbols, the Grail in this version changes form until it seems that they finally settled upon a chalice of Christ as a suitable symbol for this secret. Indeed in Wolfram von Eschenbach's Parzival, the Grail's actual keeper is a woman named "Repanse de Schoye" or "Reponse de Choix" or perhaps "*Response of Choice*"? This looks like a similar wordplay of Wolfram who named the Grail stone

---

[30] Ibid p.196

[31] Parzival, p.232

with a fake latin phrase (lapsit exillis).[32] Perhaps this name was intended for those with the eyes to see and the ears to hear what might fly over the heads of the vast majority. Repanse carrying something undescribable is suitable for an uneducated medieval audience that might never understand a mathematical beauty with cosmic significance. She represents symbolically in her name the chosen response in the symbolism created to preserve the knowledge of the Grail from those who would persecute both the knowledge and the people preserving it. It is the intellectual response to the forces of darkness or obscurantism – those who would hide an inconvenient Truth.[33] Also, the changing images of the Grail in the "Perlesvaus" may suggest that the Grail itself is something that changes form. Since King Arthur beheld all these changes, might we imagine that the historical Arthur was the head of this committee of enlightened individuals?

---

[32] See the chapter, "A Stone From Lucifer's Crown"

[33] Obscurantism – opposition to the increase and spread of knowledge.

CHAPTER 7

The Round Table

We call it the Grail because it delights the hearts of all worthy men and all those who can stay in its presence - it will not tolerate the presence of the sinful.
Robert de Boron

In the Grail stories, a poisoned spear wounded the Fisher King by piercing through his scrotum and into his thigh. The wound became gangrenous and produced a stench. Because of the wound, his "life is but a dying." However, he cannot die because he is frequently in the presence of the Grail, which keeps him unnaturally alive. The Fisher King cannot walk or sit upright and can only recline "in repose." The denial of truth is what wounds the Fisher King who represents the corpus or body of Christ, being his teaching. In symbolism, the lance or spear represents the pen because of its resemblance to the point of a feather quill that was used as a pen. (As in the likely pen name, William Shake*speare*, who shook his pen mightier than the sword at the world.) Therefore, the poisoned spear piercing the Fisher King's scrotum could be symbolic of the poisoned writing pen of the theologian, causing prejudice toward the Grail and the propagation of its secret. A virtuous knight must ask the Fisher King one of three questions to cure him. The knight must ask without being prompted, "What is the Grail for?" or, "Whom does the Grail serve?" or, he can inquire about the nature of the king's illness. The lesson here appears to be "Ask and ye shall receive," and what you may receive is esoteric knowledge. If we ask our self whom it serves, the answer may be everyone untainted by evil, as the seekers of the Grail must be open to truth. *When common wisdom denies sublime truth, the intellectual mind suffers like the Wounded Fisher King.* The Wounded Fisher King is an allusion to Christ as the Fisher of Men. He is suffering for a truth that he cannot tell to the world because of the poisoned lance that wounded him. The Fisher King cannot walk or lie down, he can only recline. Due to his pierced scrotum, he cannot reproduce, and may only discuss his wound or the Grail if specifically asked. Being able to explain the Grail heals the Fisher King and makes him, as symbolic of the body or teaching of Christ, once again whole.

Another version of the Grail story that chronologically follows the Chretien de Troyes' account is the work by Robert de Boron written between 1190 and 1199. These are a trilogy of Arthurian romances including "Joseph of Arimathea," "Merlin," and "Perceval." De Boron claims to possess secrets about the Grail that he obtained from a great book. In the text, he asks that no one shall ask him to reveal the Grail's secret, for he has decided to lie about it. This version is the first to make the Grail the cup of the Last Supper and detail how the Grail came to England through Joseph of Arimathea. This is also the first mention of Merlin the magician and the story incorporates many magical features. These include a magical chessboard that Perceval encounters on which the chess pieces move by themselves and there is a knight that charges out of a tomb to challenge Perceval. An invisible castle is filled with women who can transform into birds or old hags or beautiful women. A sole knight guards the way to this castle and in appreciation for his protection these women afford him the privilege of living with them. This scene is lampooned in the movie "Monty Python and the Holy Grail" (1974). The knight "Gawain the Chaste" (actor Eric Idle) happens upon a castle with a neon lit Grail cup appearing above its battlements. He enters the castle to find it filled with attractive young women who are starving for men and sex. His fellow knights (led by actor John Cleese), forcefully rescue him against his will in order to save him from what they call "almost certain temptation." Robert de Boron describes the Grail as something that is both wonderful, but which also some will reject, for such is the nature of the Grail. He depicts the Grail as something truly delightful! In the following quote, there is an untranslatable wordplay of Grail with "agreer" meaning, "to delight."

> We call it the Grail because it delights the hearts of all worthy men and all those who can stay in its presence - it will not tolerate the presence of the sinful.[34]

In other words, sinful people will reject the Grail and the secret knowledge it represents. Those who reject the Grail will not tolerate its presence and so may have persecuted it into the underground with their prejudice. To put it another way, if the truth symbolized by the Grail is offensive to you because it does not seem to fit within

---

[34] Robert de Boron, Nigel Bryant. *Merlin and the Grail - Joseph of Arimathea, Merlin, Perceval, The Trilogy of Arthurian Romances Attributed to Robert de Boron, Translated by Nigel Bryant.* (Cambridge, Great Britain, D.S. Brewer, 2001). p.155.

the canon of your Christian mythos, then according to De Boron your rejection of truth is sinful.

De Boron introduces to the Arthurian story the round-table and the "Siege Perilous" or Perilous Seat. The Holy Spirit instructed Joseph of Arimathea to construct a round table in the center of which he should place the Holy Grail. This is so that all who sit at the table may view the Grail equally in perspective. Only those without sin could sit at the round table and look upon the Grail. The round table has one seat reserved for Jesus himself. A false disciple named Moyse, who feels full of the Lord's grace, begs to be allowed to sit in the seat. When he does the ground opens beneath the seat, and the Earth swallows him up. De Boron weaves the round table and perilous seat from the legend of Joseph of Arimathea into the story of Percival and the Knights of Arthur. In his story the seat is reserved for the best knight rather than Christ, though the Knight must be held to the highest possible standards.

Perceval comes to Arthur's court and jousts, beating the best knights, and Arthur invites him to join the Knights at the Roundtable. There is only one seat left empty, the Siege Perilous. In ancient scholarly tradition, a seat is a title of honor, distinction, and responsibility, especially in academia. We may understand the Siege Perilous as a seat of specific knowledge. It is a "perilous seat," because to sit in the seat of the Grail Secret is to put your life in danger. Perceval asks to sit in the seat, and when he does, the rock beneath it splits with a loud crack and darkness settles over the land. A loud supernatural voice commands the knights to go on a Quest for the Grail. The honor of attaining the Grail eventually falls to Perceval, who asks the question that heals the Fisher King. After Perceval asks the question, "Whom does the Grail serve?"; the Holy Ghost descends and commands the Fisher King to give Perceval the secret of the Grail. According to the text, this is the same secret given to Joseph of Arimathea from Jesus in an earlier time. Joseph is supposed to have spent years imprisoned in a dungeon, but survived the experience due to his faith. The ghost of Jesus visited Joseph of Arimathea in the dungeon where he was imprisoned and gave him the secret of the Grail, which sustained him. Perceval learns he is a member of the Grail family possessing the secret inherited from Joseph of Arimathea. The Fisher King tells him the Lance he saw in the Grail procession is the spear of Longinus, which pierced the side of Jesus, and the Grail carried then is the vessel containing the blood of Jesus. The Fisher King lives in bliss for three days, then gives himself up to God before the Grail, and dies in peace. Perceval then becomes the new Fisher King and "Lord of the blood of our Lord Jesus Christ." Because the Grail knowledge is preserved through Perceval, the split rock under the Siege Perilous mends, again with a loud crack, and the knowledge of the Grail casts the evil enchantment out of Britain.

The New Testament does not shed much light upon the topic of the education of Jesus. The years during which he would have been educated are completely lacking in the Gospel accounts. However there are apocryphal traditions about this. One of these traditions place Jesus in the British Isles studying with the Druids. Despite the misgivings some may have over a young Jesus studying with the Druids, this idea is well within the realm of possibility. This could have occurred through that wealthy friend or relative of Jesus, Joseph of Arimathea. Joseph achieved his wealth in metals trading and shipping and he is reputed to have conducted business in the southwest of Britain. Cornwall was the chief Mining District and well known in those days for its tin. Somerset was likewise renowned for its high-quality lead. Tradition has it, that after the crucifixion, Joseph of Arimathea accompanied the apostle

Philip, Lazarus, Mary Magdalene and others on a journey to Europe. Mary Magdalene and Lazarus stayed in the south of France, and Joseph continued on to Britain. His boat landed in the Glastonbury marshes in 37 AD. Together with his companions, he climbed a nearby hill to get his bearings on the land. Legend says, he thrust his staff into the ground and announced that he and his companions were "weary all." This hill is called Wearyall Hill to this day. He bought a large tract of land (twelve hides) from the local ruler Arviragus and settled there. Joseph of Arimathea is also supposed to have brought with him the cup from the Last Supper, which became known as the "Holy Grail." This is why the Grail legends are so prominent in Britain. By some accounts, Joseph was the uncle of Mary, the mother of Jesus, and therefore would have been a great-uncle to Jesus. It is entirely possible that Joseph could have brought a young Jesus along with him on one of his business trips to the island. This is why the poet William Blake in his hymn "Jerusalem," included the lines, "*And did those feet in ancient time, walk upon England's mountains green? And was the holy Lamb of God, on England's pleasant pastures seen?*" William Blake is famous for his creative works, including a painting depicting God in a cloud with his fingers extending below into a geometer's compass. In "The Hiram Key Revisited," by Christopher Knight and Alan Butler, they claim there is little doubt Blake was a Freemason, "since many of his most famous paintings display such powerful Masonic imagery."[35] They claim that William Blake was well known for membership in the Ancient Order of Druids (*Druidh Uileach Braithrearchas*) formed originally at a gathering of prominent Freemasons on the autumn equinox in 1717, writing "In 1796, the Druids chose a new Grand Master:

[35] Christopher Knight, Alan Butler, *The Hiram Key Revisited, Freemasonry: A Plan of a New World Order.* (New York: Metro Books 2010). p.207.

William Blake, one of the most remarkable creative personalities of this, or indeed any other, period of English history." Also in "The Hiram Key Revisited," they claim the Scottish King James VI, who became the king of Britain, Scotland, and Ireland on the death of Queen Elizabeth (as James 1), and who commissioned the English translation of the Bible called the King James Bible, *was also a Freemason,* for "The records of the Lodge of Scoon [*sic*] and Perth No. 3, which date back to 1598 state that King James VI was made a Freemason on 15 April 1601."[36] Of interest they note that King James introduced the Scottish sport of golfing to the English in Blackheath, and that golf was used as a cover for Masonic activities; which made eavesdropping difficult. King James is considered one of the most intellectually capable of British monarchs in history.

This symbolism of the splitting stone is explained by De Boron's tale of Joseph of Arimathea. During the crucifixion, Joseph observed a drop of blood falling from Jesus that fell on a stone at the base of the cross. The drop of blood caused the stone to split apart. When Joseph takes Christ off the cross, he sees his wounds still bleeding. Recognizing the blood of Christ has powerful magical properties; he captures some of the blood into the cup of the Last Supper. The split stone is a reference to the rock of Matthew 16:18, "on this rock I will build my church." The stone at the base of the cross is symbolic of a foundation stone of the church and the drop of blood is symbolic of some knowledge that fractured or created a schism in the Church. Perceval as scion of Joseph of Arimathea is a drop of the bloodline that preserves the same knowledge that caused the rock under the cross to split. Therefore, when Perceval sits in the Perilous Seat, he

[36] Ibid. p.187.

is like the same drop of blood that split the stone under the cross. However, the same knowledge that splits the stone also restores the stone to wholeness at the end of the story. The symbolism is clear. The wounded Fisher King represents those who preserve some knowledge of Christ who are symbolically wounded by a poisoned pen (spear) that renders them both sterile and unable to walk out into the world. The enemies of this knowledge made it very dangerous to "reproduce," like the Fisher King, or *to publish this knowledge.* Thus, it can only be stored safely, "but in repose," like the Fisher King who cannot walk. Like scientific knowledge in the Dark Ages, it could not advance out into the world. Further, the Grail "Quest" to hunt or seek out Truth is antithetical to dogmatic theologies that encourage blind faith and obedience. To go on a quest is to "question" the veracity of what authority has told you. To heal the wound of the Fisher King, Perceval must ask him, "Whom does the Grail serve?" *One thing is clear from this text, if it offends you and you cannot tolerate its presence, then for certain, it does not serve you.*

There is a real historic Round Table hanging in the Great Hall of Winchester Castle in Hampshire England. Although by legend it is supposed to be King Arthur's table; in 1976 it was examined by scientists and by tree ring evidence and carbon dating; it probably was made in about 1290 AD. According to John Harding in the "Chronicle of England" (1463), "The Rounde Table at Wynchestere beganne, and ther it ende, and ther it hangeth yet." At the center of this table is a Tudor Rose, whose name and origin is with the Tudor Dynasty. There is fertile ground here for research, because here is a perfect example about how the truth may be hidden in plain sight for all with "eyes to see". The Tudor Dynasty Rose with barbs is depicted in the center of this very real historic round table in the Great Hall of Winchester in

Hampshire England. It should be pointed out that a rose is not a depiction of the Grail in any medieval grail romance.

Figure 2 Tudor Dynasty Barbed Rose

CHAPTER 8

Masonic Secrets

But we speak the wisdom of God in a mystery,
even the hidden [occult] wisdom,
which God ordained before the world unto our glory.
1 Cor 2:7 KJV

In Parzival, Wolfram von Eschenbach emphasizes that the guardians of the Grail are Templar Knights. This is despite the fact that the Templar order was not established until centuries after the time of King Arthur. Authors have pointed this out as a glaring anachronism or use of poetic license in projecting his contemporaries, the Knights Templar, into his Grail story. The King of France horribly persecuted this famous monastic order of Templar Knights and accused them of terrible heresies in 1307. The Pope officially disbanded their order by decree in 1312. The Order went underground and there is much evidence it resurfaced in the ancient institution of Freemasonry that exists to this day.[37] Freemasons proclaim a continuity of tradition back

[37] John J. Robison, *Born in Blood - The Lost Secrets of Freemasonry*, (New York, NY: M. Evans & Company, 1989).

to the time of King Solomon and ancient Egypt. Evidence of this is in the extreme antiquity of their symbols. The symbols of the Masons, such as the compass and square, have been found under the base of ancient Egyptian obelisks, which gives their claims credibility. If this ancient pedigree is true, then it is valid to place the Templar Knights in the time of King Arthur. The founding fathers of America were famously Freemasons, such as George Washington and Benjamin Franklin. Although some think it is a sinister organization, Freemasonry is really a form of hidden college preserving ancient wisdom.

Wolfram declares that his version of the Grail story is correct and that the Chretien de Troyes account is merely a fantastic fable. He also considers his work to be a form of "initiation document" implying knowledge of esoteric secret societies. Wolfram is very clear that the Grail is something of immense consequence, and repeatedly he drops hints and encourages the audience to read between the lines. Wolfram's knowledge of the Grail comes indirectly from the astronomer Flegetanis who is described as half Hebrew and half heathen. If the Grail is so important in Christian lore, then why should it come to Wolfram through Kyot from Flegetanis in Muslim Spain? The answer to this is in the symbolism of the neutral angels, which could represent esoteric societies transcending all faiths.

Scholars have spent considerable energy debating whether Kyot and Flegetanis are real or fictitious. Historians have established that Kyot de Provence would almost certainly have been Guiot de Provins who was known to officially represent the Templar's. He was a contemporary of Wolfram and a troubadour and monk who did live in Provence. We know that Guiot attended a chivalric festival of Pentecost in Mayence, Germany in

1184 and it is quite possible Wolfram could have met him there. Learned men were rare in those days and would have sought each other out. If Kyot is real, then we may imagine that Flegetanis the astronomer was also a real person as well. Wolfram connects the Templars and the Grail family to esoteric society through a unique service that they provide in addition to Guarding and preserving the Grail. Wolfram writes:

> The Grail selects only noble company. Knights, devout and good, are chosen to guard it. The coming of the high planets brings this people great sorrow, young and old alike. God's anger at them has lasted all too long. When shall they ever say yes to joy? [...] I will tell you something more, whose truth you may well believe. Fortune often faces them with win-and-lose. [...] And if anywhere a land loses its lord, if the people there acknowledge the Hand of God, and seek a new lord from the Grail Company, they are granted one. They must treat him with courtesy, for the blessing of God protects him.[38]

The Grail family calls young men and women into its service from the outside world. They are initiated into the mystery of the Grail. Young maidens are appointed to care for the Grail itself. Young men that come to the Grail become chaste knights who defend the Grail. Wolfram is very explicit that the knights must forego woman's love, and only the Grail King, or those men selected to be lord of other lands might wed. The Grail family sends specially trained men out into the world on specific missions - and sometimes to occupy a throne. *The Grail has the power to make kings.* If some people lose their lord, they may request one from the Grail Company and they will provide a man who is wise and just in governing. However, those people must never question his name, origin, or lineage, for if they do they will have his help no longer. Wolfram

---

[38] Eschenbach, *Parzival,* p.251.

claims this is the meaning of the legend of Lohengrin, the Swan King, who comes to a people across a lake in a boat pulled by a Swan and restores justice to the land. When he is asked about his identity he leaves the people and sails back across the lake. Wolfram emphasizes two points about this at the end of the story. One is that these appointed rulers are Templar Knights and the second is that they will restore the people's human rights. It seems extraordinary in the context of a twelfth century popular fiction to discuss human rights that we might tend to think of as a particularly modern concern.

From the passage quoted above, it is clear that at some point in the past the Grail family incurred God's wrath, "God's anger at them has lasted all too long. When shall they ever say yes to joy?" Whatever curse they formally suffered, at least in the time of Parzival they enjoy divine favor and have great power. We may glean a great deal from these passages. Let us start with a supposition that the Grail Company is an esoteric society possessing primarily a treasure of esoteric knowledge, of which the Grail is a specifically potent and forbidden secret. The society provides for a way to perpetuate itself by receiving young blood into its company, and provides a crucial service to the world. Should a land lose its lord or succumb to despotism or tyranny that tramples human rights, the people "in the know" about the esoteric society of the Grail Company, can request their assistance to restore balance and human rights to their land. Since this may involve overthrowing a tyrant clinging to power by resorting to evils, it is understandable that, "Fortune often faces them with win-and-lose." In such a risky business, we may well imagine that the Knights Templar, the virtuous and chaste warrior monks, would be a valuable asset.

This function of the Grail Company sounds very similar to the Freemasonry of the American founding fathers. The American government was built upon firm foundations designed to preserve human rights. The Freemasons are famously neutral regarding issues of religion, since they know that religious persecutions were some of the greatest evils the world has ever known. To become a Mason you only need to acknowledge a creative intelligence in the universe, and Masons will never require you to change your faith, whatever it may be. This is identical to the quote above in which people needful of assistance from the Grail community need only, "acknowledge the hand of God." This acknowledgment does not require adherence to a specific creed. The esoteric society of the Freemasons also functions like the Grail Company in providing politicians and advisers that stand for justice. Freemasons have taken the office of the president of the United States at least 15 times, starting with George Washington and including Franklin D. Roosevelt and Harry S. Truman.

The fact that the Grail family has incurred God's wrath is because they are inheritors of the mythical legacy of the neutral angels that are said to have brought the Grail to earth during the time of the angelic rebellion. A dogmatic patriarchal culture may promote absolutist philosophy and in Revelation 3:15-16 God says, "I know your works, you are neither hot nor cold [...] so then, because you are lukewarm [...] I will spew thee out of my mouth." Thus, if you are neutral, you are the enemy. This type of thinking is found in all absolutist and fascist philosophies. When these philosophies become prominent people of neutral stance are punished. This happened in Germany, where Nazis persecuted, jailed, and in many cases executed tens of thousands of Freemasons.

As expressed by followers of absolutist philosophy "God's Anger," still exists against the "neutral angels" to this day. Esoteric societies, such as the Freemasons, are still routinely slandered although this is usually by ignorant people dead to the idea of esotericism. For instance, some Christian fundamentalists have slandered the Freemasons as being a "satanic organization." Freemasons do little to discourage this. Like true knights who can defend themselves, they allow themselves to be the target of such slander rather than as in the past, when innocents would be persecuted. Masonic belief in freedom of religion in many previous ages would be a cause of excommunication from the Catholic Church. Perhaps because of this reality, at the end of his book Wolfram retracts his suggestion that God may forgive the neutral angels.

> Let me atone for my error [...] you heard from me that the banished Angels were at the Grail with God's full support till they should be received back into His Grace. But God is constant in such matters: He never ceases to war against those whom I named to you as forgiven. Whoever desires to have reward from God must be in feud with those Angels. For they are eternally damned and choose their own perdition.[39]

Despite the happy conclusion in Parzival, it would seem the Grail keepers must always be cautious of "God's wrath" from people who espouse intolerant belief systems. By knowing too much Truth, the neutral Grail Keepers cannot descend into absolutist thinking, because they know that Truth has two sides like a coin. The summary of the mission of the neutral angels is that they preserve knowledge that may in some ages be controversial and persecuted. Doing so marks them as

---

[39] Ibid, p.396.

being perversely disobedient to God. In Freemasonry, there are thirty-two published degrees of study. Might it not stand to reason that if the Freemasons are the spiritual inheritors of the Knights Templar that their path of initiation would be like a Grail Quest? Imagine you were a sleuth following a trail of clues leading from the Grail to the Templars and then to the Freemasons. Might you be reasonable to speculate the secret of the Grail would lay in repose, like the Fisher King, amid the underground teachings of the Freemasons or other secret societies? In particular, is it possible that the secret of the Grail is revealed to those who look for it in the highest degree called, "The Royal Secret"?

CHAPTER 9

Deprecating Holy Blood, Holy Grail

But Christ loved her more than all the disciples and used to kiss her often on the mouth.
Gnostic Gospel of Philip

The inspiration for this book was a dissatisfaction and resulting fascination with earlier Grail theories such as the one presented in the bestseller, "Holy Blood, Holy Grail" (1982, 1983) and the large number of books which capitalized on the popular theory about Jesus bearing a child and establishing a royal bloodline. This is an intriguing theory, and it inspired a great mystery novel called "The DaVinci Code" (2003) and a resulting popular movie of the same name in 2006. There was however, a glaring problem in the original book. That problem was that after discussing Wolfram von Eschenbach's "Parzival", and especially the astronomer Flegetanis unique to Parzival, the authors present a synthesis of all the evidence and come to a conclusion which in no way explains the role of an astronomer in their theory. Also there was a discomfort with a theory of the significance of a bloodline, for several reasons. One is that in the modern world it is realized that DNA does not propagate true knowledge or character to merit an individual worthy to serve in a role of authority. Along this same reasoning there is an esoteric and Platonic idea that the things of this world are impermanent, while abstracts such as ideas and principles of philosophy have a far greater permanence because, like geometry and mathematics and logic, they have an existence independent of this world. Ideas are in this world, but not necessarily of this world.

One famous painting that is supposed to contain encoded Grail symbolism is Leonardo DaVinci's famous fresco, “The Last Supper.” This depicts the supper at which Jesus initiated the sacrament of Holy Communion that involves the partaking of bread and wine. “The Last Supper” depicts the definitive moment for the Holy Grail as a cup and yet DaVinci chose to leave it out of the picture. Rather than having a large “Grail” to pass around, there are instead thirteen cups without stems. Authors have declared that the missing Grail cup is highly significant since the Eucharist (the mystical drinking of Christ’s blood and eating his body through bread and wine) is generally considered the most significant event to depict from the Last Supper. Actually, there is nothing mysterious here. DaVinci is just depicting the most detailed account of the Last Supper from the Gospel of John. This account is notable because, unlike Matthew, Mark, and Luke, there is no Eucharist mentioned! It is also by far the longest version, spanning five chapters.[40] Also note, the figure to the right of Jesus is a departure from the Gospels. This is supposed to be St. John, known as “the

---

[40] John 13-17.

beloved," but the figure to his right is a woman. There is no ambiguity or androgyny in any of the figures depicted in this painting. All the other figures look manly, and this figure is clearly female. They are clothed as mirror images with inverse colors suggesting a harmonious reciprocal relationship. Jesus wore a red robe and blue cloak; the woman wears a blue robe and a red cloak. If you take the image of the woman and transpose it to the opposite side of Jesus then her head will rest exactly on his shoulder with her body molded to his side.

In the early Christian Church, and particularly among the Gnostics, the Magdalene played a much more important role. The early church gave her the title of "Apostola Apostolorum" or "The First Apostle," because she was the first to see the risen Christ. However, the Gnostic texts go way beyond this to proclaim her importance. She is not merely important among the woman, she is important above all the men as well. It appears that she was the bridge between Jesus and all his disciples and she interpreted and explained his words to their benefit. According to the Gnostic texts, Peter was antagonistic toward Mary and misogynistic. In the Gnostic Gospel "Pistis Sophia," Mary Magdalene says, "Peter makes me hesitate: I am afraid of him, because he hates the female race." In the Gospel of Thomas, Peter says, "Let Mary leave us, for woman are not worthy of life." Furthermore, according to the Gnostic gospels, the relationship between Jesus and Mary was not purely platonic. For example, in the Gnostic Gospel of Philip:

> But Christ loved her more than all the disciples and used to kiss her often on her mouth. The rest of the disciples were

offended by it and expressed disapproval. They said to him, 'Why do you love her more than all of us?' [41]

In the same Gospel, Mary Magdalen is called the companion of Jesus. While today the word "companion," means friend or associate in a purely platonic sense, the original Greek word actually meant consort or sexual partner. If Mary Magdalene were really his lover, then her enigmatic importance in the Gospels would be clarified. Nevertheless, a large percentage of modern Christians react to this idea with moral indignation. Jesus as exemplar of the "perfect man," is supposed to be celibate. However, he was called Rabbi and within the Hebrew culture of Jesus' day, rabbis were always married. In the Judaic tradition, celibacy was then and is today, considered to be actually sinful. A tradition attributed to a near contemporary of Jesus, Eliezer Ben-Asau (ca. 90) is the saying that, "Whoever does not engage in procreation is like someone who spills blood."[42] If Jesus were a bachelor and promoting celibacy it should have been cause for comment, yet specific mention of it is lacking in the Gospels. Yet if Jesus were intimate with Magdalene and not married, then this behavior might justify the comment of John 9:24, "We know this man to be a sinner." Yet this idea; indeed the entire premise of "Holy Blood, Holy Grail" is a gross simplification of the matter.

The most common conclusion today is that the Grail legends refer to Mary Magdalene as the embodiment of the Grail. According to some accounts, she was supposed to have given birth to a daughter of Jesus named Sarah, raised in the south of France. By bearing Jesus a daughter, the Magdalene would literally be the vessel containing the

---

[41] Willis Barnstone, *The Other Bible.* (New York, Harper Collins, 1984). p.92.

[42] Uta Ranke-Heineman, *Eunuchs For the Kingdom of Heaver – Women, Sexuality, and the Catholic Church.* (New York: Doubleday, 1990). p.40.

blood of Jesus or his bloodline. The term "Holy Grail" may come from "San Greal," which comes from "Sangreal," or "San Real" which means "Royal Blood." The tradition about a child named Sarah is conjecture on traditions from the south of France. The scholar Margaret Starbird wrote an account of this in her book "The Woman With the Alabaster Jar."[43] In the small coastal French town of Les Saintes-Maries-de-la-Mer, there is a festival every May 23-25 at a shrine in honor of Saint Sarah the Egyptian. This festival honors an Egyptian child that accompanied Mary Magdalen, Martha, and Lazarus who landed in a small boat at this location in about 42 A.D. If she were a child of Jesus born to Mary after her escape to Alexandria following the crucifixion, then she would have been about twelve when arriving in south France.

Some have suggested the Magdalene was of royal descent of the tribe of Benjamin or possibly a high-ranking priestess of the cult of Isis. The gospels support that Jesus was of the House of David and descended from King Solomon. If married this may have been a dynastic marriage and their descendents may have spawned a royal line, such as the Merovingian dynasty in France. Books such as "Holy Blood, Holy Grail," take this as a primary premise for the existence of secret societies such as the alleged Priory of Sion. Supposedly, they would have preserved from destruction documents tracing this lineage and protected the lives of these people from possible assassins.

Since the world today has moved towards democracy and away from monarchy, it is hard to imagine anyone being overly impressed with bloodline and lineage. The time of Jesus was about one hundred generations back. Anyone who has tried to trace his or her lineage back even

---

[43] Margaret Starbird, *The Woman With the Alabaster Jar.* (Rochester, Vermont: Bear & Company, 1993). p.60-61.

a few generations knows that it quickly becomes an exponentially large problem. In "The Seven Mysteries of Life," (1978) Guy Murchie wrote about humanity's common ancestry.

> Although few of his members seem to realize it, man's relation to himself is fairly easy to measure and is surprisingly close. In fact, no human being (of any race) can be less closely related to any other human than approximately fiftieth cousin, and most of us (no matter what color our neighbors) are a lot closer. Indeed this low magnitude for the lineal compass of mankind is accepted by the leading geneticists I have consulted (from J. B. S. Haldane to Theodosius Dobzhansky to Sir Julian Huxley), and it means simply that the family trees of all of us, of whatever origin or trait, must meet and merge into one genetic tree of all humanity by the time they have spread into our ancestries for about fifty generations.[44]

A generation is generally about twenty to twenty-five years. Therefore, all people on Earth have a common ancestry of those who were alive in about 750-1000 C.E. We all have two parents, four grandparents, and eight great-grandparents. For each generation our number of ancestors grows by a power of two. Thus within only thirty generations our number of ancestors would exceed one billion, yet this number is much larger than the entire world population of that time. Once you go back several generations, you will find the same ancestors showing up in multiple branches of genealogies. The further back in time you go in the genealogies of any two people, the common ancestors will occur with increasing frequency until eventually all ancestors are the same. This means all humans are cousins related by blood of an unknown degree highly unlikely to be a large number, and probably not more than fifty. *Therefore, if Jesus had a child, or even if*

---

[44] Guy Murchie, *The Seven Mysteries of Life*. (New York: Houghton Mifflin, 1978). p.345.

he did not, we all have the blood of the House of David running through our veins!

CHAPTER 10

December 25th as Gnostic Pleroma

That which is called the Christian Religion existed among the ancients and never did not exist, from the beginning of the human race until Christ came in the flesh, at which time the true religion which already existed began to be called Christianity.
St. Augustine ca 410 CE[45]

In the 17th century James Usher (1581-1656) the Archbishop of Armagh (Church of Ireland), dated the creation of the world by a literal reading of the Bible, to nightfall on Saturday, October 22, 4004 BC. This date for the age of the world is still respected for its biblical scholarship. More recent archaeological evidence corresponds 4,000 BC to the founding of the first ancient Sumerian towns in the Fertile Crescent region near the Tigris and Euphrates rivers.[46] This was when mankind

---

[45] Referenced as *Retractions 1.13.3*, also *Sermo 10, De Santis.*

[46] The archeological interpretations in this chapter are exegetically from Joseph Campbell in "Flight of the Wild Gander".

transitioned from Neolithic culture to agriculturally based civilization. All the high cultures of the world have inherited from what happened there. By 3,500 BC, the Sumerian cities of Ur, Kish, Lagash, Eridu, Sippar, Shurappak, Nippur, and Erech had been established. A king and queen or princess governed these cities, and the rulers were identified with a god and goddess. Ancient Sumer at this time also had, for the first time in history, temples that grew in importance over time and became centers of politics, religion, and economics. At a date that can be almost precisely fixed at 3,200 BC, in the archaeological stratum known as Uruk B, there was a sudden synergy. Newly minted elements of civilization came together to form a new whole that was greater than the parts. Civilization as we now know it, bloomed and flourished. These inventions of civilization included the first system of writing and the literal beginning of documented history. The wheel appears at this time, as does musical harmony. There were established rituals in human life. A mathematically structured calendar regulated the seasons of life. At this time two numerical systems that are still used to this day were developed. The first of these is the decimal system, based on ten fingers or digits, which was used for accounting business. Ancient Sumer produced the first written accounts of money exchange, loans made, and interest paid. The second numerical system invented was the sexagesimal, based on the number sixty, which is used for measuring space and time to this day. We méasure the hour in sixty minutes and we measure our minutes in sixty seconds. Our circle is still measured in three hundred and sixty degrees (6 x 60). Three hundred and sixty days plus five (365) measure our year.

The ancient Sumerians had invented two of the "magical pillars" of civilization. One of these pillars is economics and the other is organized religion. Money and the creation of wealth involve the societal acceptance of "transubstantiation" which is the offering of one substance in the place of another. Money is an abstract or magical substance that we exchange for material goods. It is essential for modern civilization to function and is essentially "conjured" out of thin air. In her book, "Solomon's Treasure," Tracy Twyman writes:

> The power of money to transform almost any thing or situation into another is similar to the alchemical power of the so-called "universal agent" or "Philosopher's Stone", and the act of turning paper into dollars is like turning lead into gold. The members of the Federal Reserve Board are in many ways like sorcerers, conjuring wealth seemingly out of thin air and distributing it at will to transform the American economy according to their desires. The dollar is "fiat currency", declared into existence by the central bank in a manner similar to the creation of the universe by the divine words "Fiat Lux" – "Let there be light!" [47]

Money is an abstract imitation of organic nature. Money grows through compounded interest and the average yearly growth rate of investing is similar to that of trees in nature and dividends are like a fruit plucked thereon. The second of these magical pillars is organized religion, which has been recognized as the single most effective means of controlling the masses of humanity that was ever devised. This was a key to large numbers of people living peacefully in the close quarters of cities.

In the Neolithic cultures preceding ancient Sumer, and in the Neolithic cultures that still exist today, most humans could learn everything they needed to function in their society by the time they reached puberty. As civilization became more complex, it became impossible

---

[47] Tracy R. Twynan, *Soloman's Treasure – The Magic and Mystery of America's Money*. (Portland Oregon: Dragon Key Press 2005). p.14.

for individuals to know everything about the society in which they were raised. This led to specialization of education and increasingly longer times before children became adults and ready to participate in the culture. In the "Flight of the Wild Gander," Joseph Campbell writes that civilization originated in extending infancy by surrounding individuals with a mythology designed to provide a sense of security. This is why the rational analysis of mythology (religion) may elicit puerile responses from individuals, no matter what their age. Joseph Campbell compares the function of myth in society to be like the marsupial pouch. Marsupials represent an evolutionary intermediate stage between monotremes like the spiny anteater of Australia, whose progeny are born from eggs, and the placentals, whose young require a longer gestation period in the mother. Man is biologically a placental whose gestation over time has become increasingly inadequate. The Homo Sapien requires years before it may forage for food on its own and as many as twenty years before becoming an adult. Joseph Campbell writes:

> Society, as a fostering organ, is thus a kind of exterior "second womb," wherein the postnatal stages of man's long gestation – much longer than that of any other placental – are supported and defended. [...] Rites, then, together with the mythologies that support them, constitute the second womb, the matrix of the postnatal gestation of the placental Homo Sapiens.[48]

What builds this second womb is a "tight-knitting of defensive fantasy," with the external reality of our society. Many people stay in this defensive fantasy their entire lives. This protects the individual psyches from their fears

---

[48] Joseph Campbell, *The Flight of the Wild Gander – Explorations in the Mythological Dimension.* (South Bend, Indiana: Regnery/Gateway Inc. 1951, 1969) pgs.53,55.

of what James Joyce called the, "grave and constant in human suffering." Religion throughout history has provided human beings with the earliest form of "insurance" against our deepest fears. This was part of the evolution of man and organized religion.

The organization of Sumerian culture was a synergy between new inventions like writing, mathematics, and astronomy, along with the older ones such as mythology. The same basic culture spread over time all over the world. It inspired the First Dynasty of Egypt c. 2,800 BC. It reached Crete and the Indus Valley c. 2,600 BC. According to Joseph Campbell the inspiration of the culture was a "coordinating principle" which was an "image of man's destiny as an organ participating in the organization of the Universe," via an "all-suffusing, all-informing principle or energy." After many thousands of years of humankind's striving upward to evolve into a greater being, humanity learned our position within a greater vision of the Cosmos. At the dawn of modern civilization in about 4000-3200 BC, humanity in Sumer attained a sort of epiphany that fulfilled a long sought for mass psychological understanding. According to Joseph Campbell this sociological need:

> [...] must have been fulfilled with the recognition, some time in the fourth millennium B.C., of the orderly round-dance of the five visible planets and the sun and moon through the constellations of the zodiac. And this celestial order was to become for all the civilizations and philosophers of the world the model of the revelation of destiny.[49]

The scientific revelation of the heliocentric order in the heavens above was reflected below in the Sumerian cities. The cities were designed like a mandala of the solar system

[49] Campbell, *Flight of the Wild Gander.* p.154

centered around the temple and the king as the solar deity on earth and the people in orbit around the sacred center. The solar system was the Cosmos above; the city was the middle cosmos or mesocosm between the individual and the divine. The individual was the microcosm reflecting the divine heavens within. The great mystery was expressed in the perfection of the circle in three hundred and sixty degrees and the mystery of the reality of Earth's orbit around the Sun in three hundred and sixty five days of the cycle of the year. The five days difference had a mystical meaning that could only be understood *in a precise knowledge of the orbital harmonies of the planets.* According to Joseph Campbell:

> And those five intercalated days – which represent the opening through which spiritual energy flows into the sphere of time from the pleroma of eternity, and which, consequently, are days of feast and festival – coorespond in the temporal mandala to that mystical point in the center of the spacial mandala which is the sanctuary of the temple, where the earthly and heavenly powers join. The four sides of the temple tower, oriented to the four points of the compass, come together at this fifth point, where the energy of the pleroma enters time – and so once again we have the number five added to three hundred and sixty to symbolize the mystery of the immanence of eternity in time.[50]

The Pleroma survives as a Gnostic concept of the *originating point* from which the divine emanates into our own existence. It is called, "the light existing above our world." Astronomical truth, was wedded together with ancient religion, and civilization as we know it was born!

In ancient Assyria the two symbols for Venus/Ishtar (the goddess) were an eight-pointed star and a five pointed star – which as will be shown, is proof that 5,200

---

[50] Ibid, p.153.

years ago the ancient Assyrians understood the heliocentric (Copernican) cosmology so bitterly persecuted by the medieval Church for ~1,200 years. The five intercalated days of feast and festival are celebrated between December 25th and the New Year. Christianity adopted this practice from long standing ancient tradition from pagan times, celebrating the birth of the new winter Sun or Son of man. According to Kersey Graves, "When we find that the ancient pagans were in the habit of celebrating this venerated twenty-fifth of December as the birthday of their Gods in the same manner Christians now celebrate it as the birthday of Christ, we are driven to admit that something more than mere fortuitous accident must be adduced to account for the coincidence."[51] The mystery of the Pleroma is part of the explanation of why there was a pre-Christian pagan tradition of the savior's birth being celebrated on December 25th. The Pleroma is an aspect of the Grail secret as the point of the year when its energy enters time and which is acknowledged as a sacred time in very ancient tradition. The secret of the Grail includes astronomical knowledge with a mystical revelation about life entering into the Cosmos. This is the same as the mysterious concept of the Pleroma through time, whose celebration since the beginning of civilization as we know it, has begun on December 25th to mark the end of one cycle of the Earth around the sun, and to begin the regeneration of life for the next.

---

[51] Kersey Graves, *The World's Sixteen Crucified Saviors, or Christianity Before Christ*. (Montana, USA: Kessinger Publishing Co.) p.68.

CHAPTER 11

English Alphabet as Pentacle

Hear the word of the Lord, O nations,
And declare it in the isles afar off, and say,
"He who scattered Israel will gather him,
And keep him as a shepherd does his flock."
Jer 31:10 NKJV

Among those who study the evolution of languages, the English language is an enigma. This is because the English language evolved extremely rapidly under special circumstances. English is divided into three broad styles associated with historical periods. These are Old English, Middle English, and Modern English. Old English, which is the language that "Beowulf" was written in, is almost unintelligible to the modern ear. Yet, within the time span of only about one generation, Old English evolved into Middle English, which is easily understood by modern people. The traditional explanation for this is the extraordinary circumstance of the time of change. In the study of philology, which is linguistics and its evolution, the extremely fast evolution of Middle English is acknowledged as significant, but you might not expect that pursuing the meaning of this was a path that would lead eventually to *a major clue to solve the mystery of the Holy Grail.*

The Middle English language evolved in the historical period following the Norman Conquest of Britain in 1066 A.D. As the Normans assimilated themselves into Britain's population, the French language of the Normans and the Old English of the Britons merged to form a significantly new language. This might be enough to explain the rapid development of the English language, but for the following. In the generation following the Norman Conquest, the English alphabet was willfully created. According to William Eisen, "The [English] alphabet is essentially the 23-letter classical Latin alphabet of yesterday, but with three additional letters, the J, U, and W, which were added around the 11th century A.D."[52] They adopted the Roman alphabet and added three new letters to it - JUW. The letters phonetically spell "Jew" and this seems to imply a Hebrew influence. History records little about the Norman people. They came to the north of France from Scandinavia and settled there with permission of the French king and as a group adopted the French language as their own. They had an advanced architecture and legal system. It has been conjectured that the Normans might have been one of the lost tribes of Israel, such as the tribe of Benjamin. One such book about this is "Israel-Britain or Anglo-Saxon Israel," by Adam Rutherford. This book also claims the Royal House of Britain is descended from King David of the Old

[52] William Eisen, *The Universal Language of Cabalah.* (Marina Del Rey, CA: DeVorss & Company 1989) p.21.

Testament and provides one hundred and twenty-three names of this lineage.[53]

The Scottish people also claim to be descended from the Hebrews. The "Stone of Scone," which may be seen today in Edinburgh Castle, is the famous coronation stone upon which the Scottish Kings and British Monarchs made their oaths for many centuries. It is also known as the Stone of Destiny and the Coronation Stone. It is believed to have been the same stone upon which Jacob laid his head and dreamt of the ladder to heaven upon which the angels could ascend and descend to earth as described in Genesis chapter twenty-eight. In 1320 A.D., King Robert the Bruce and his noblemen signed the "Declaration of Arbroath," which traces their migrations as a people back to the time of Exodus, when the Jews crossed the Red Sea from Egypt, and up to the time when they settled Scotland some twelve hundred years later. This document is also informally known as "The Scottish Declaration of Independence." It contains the sentence, "It is in truth not for glory, nor riches, nor honours that we are fighting, but for freedom – for that alone, which no honest man gives up but with life itself."[54] This declaration was the culmination of the efforts of a generation. Robert the Bruce finished the fighting begun by William Wallace, who was executed by King Edward I of England August 24th 1305. Robert the Bruce picked up the cause of Scottish independence after Wallace's death. Historically coincident with this time on Friday October 13th 1307, the Knights Templar were brutally suppressed and subsequently tortured for heresy across Europe. This is the origin of the tradition of Friday the 13th being an

---

[53] Adam Rutherford, *Israel-Britain or Anglo-Saxon Israel – An Explanation of the Origin, Function and Destiny of the Anglo-Celto-Saxon Race in the British Empire and U.S.A.* (Muskogee, OK: Artisan Publishers 1934, 2002) pps.64-66.

[54] Capt, E. Raymond, *The Scottish Declaration of Independence.* (Muskogee, OK: Artisan Publishers, 1983) p.42.

unlucky day. There is a legend that an underground fighting force of Templar Knights helped Robert the Bruce turn the tide against the English on June 24th 1314 at the famous Battle of Bannockburn. Graveyards of that generation in Scotland have Templar symbols on gravestones such as the *Scull and Crossbones* used on Templar battle flags.[55] This indicates that some Templars escaped persecution and went underground in Scotland and intermarried and fought with the Scots. It was here and during this time that Scottish Freemasonry was established and which maintained secrecy before re-emerging publicly in London in 1717, as told in detail by John J. Robison, in his book "Born in Blood."[56] Further research may reveal that the Scottish scholar Duns Scotus had ties to the Templars and Freemasons as a history of Duns Scotus declares that he left Paris suddenly in October 1307 for reasons unknown. It is from Scotus that we have the name of the conical "Dunce Cap" used to shame students into studying. The conical hat is a symbol of wizardry harking back to the time of the Alexandrian scholars and their work on the geometry of the conic sections. It is said he encouraged wearing a conical hat as a "thinking cap" to encourage intellectualism. Duns Scotus died unexpectedly in November 1308 during the time of the Templar persecutions. One might conjecture his death being a result of associations with the Templars.

The big picture of history indicates that following the Norman Conquest in 1066 the English language and alphabet were created with a special purpose. This

---

[55] The entire Templar fleet of ships escaped from France, presumably with the Templar treasure that was never found. Although the topic is outside the scope of this book, it is none the less extremely significant that the Scull and Crossbone flag was used not just by Templars, but also by pirates in a later era. Masonic initiation contains the words, "You will become a friend to pirates."

[56] John J. Robison, *Born in Blood – The Lost Secrets of Freemasonry*. (New York, NY: M. Evans and Company, 1989).

indicates the Norman people possessed advanced esoteric knowledge and a mission. That the Normans possessed esoteric knowledge is evident by a visit to the Philadelphia U.S.A. Masonic Grand Lodge, which features Norman architecture in its external structure and has a hall devoted to the Normans.[57] The Norman architecture being celebrated in a building rooted in Scottish Freemasonry indicates some ancient cooperation and understanding between the Normans and the Scottish. The Normans had very long-term plans extending over centuries of time and if they were one of the lost tribes of Israel, they may have evolved an intense warrior culture with the intent of regaining Jerusalem. They may have sought to fulfill the prophecies of Jeremiah that the lost tribes would return to Israel and reclaim the land. "Hear the word of the Lord, O nations, And declare it in the isles afar off, and say, 'He who scattered Israel will gather him, And keep him as a shepherd does his flock.'" (Jer 31:10 NKJV) The mention of the "isles afar off" in Jeremiah, are believed to refer to the British Isles far from Israel. The supposed prophetic command of God (actually the wishful prophecy of Jeremiah) to gather his people there and return to Israel explains the Norman invasion of England and the Crusades in the Holy Land in the following centuries. All of this history has been given to provide an historical context for the time in which the English language and alphabet developed. It may be revealed to have been created willfully rather than organically as linguistic science or philology might claim. *The remainder of this chapter may be considered proof of this conjecture.*

Because of the addition of the new letters JUW in specific placement, the English alphabet contains two very significant mathematical constants encoded within it.

[57] This can be seen over the internet at pagrandlodge.org.

These are the values of Pi and Phi. It should be obvious that having two mathematical constants encoded cannot possibly be by chance; it is by design to send an encoded message. The Pi ratio (3.1416) is the ratio of the circumference of a circle to its diameter. The Phi ratio (1.6181) is a ratio that is found abundantly in life on earth. These ratios would not be so encoded in the English alphabet without the JUW letters added. You may reveal the encoded Pi and Phi ratios by arranging the characters in a circle and separating the symmetrical characters from the unsymmetrical ones.[58] (See the Appendix for details.) The fact that the Pi/Phi values are encoded in the English alphabet is evidence that the people who created the alphabet were very clever and believed that these mathematical constants were very important. It is well known that Pi is significant in science and architecture, but it may not be obvious to you what significance Phi may have. The question to ask is, why did they go to so much trouble to encode Pi and Phi? Having the ratio of Pi encodes a circle in the English alphabet. Phi exists infinitely in the pentagram or five-pointed geometric star.

Combined together the circle and pentagram make the pentacle that is associated with witchcraft and which was persecuted in the Dark Ages. It is to this day associated with magic and sorcery. The Divine Ratio figures prominently in artwork and architecture, and is called divine, because it is a ratio or proportion, which is found everywhere in life forms on Earth. Leonardo DaVinci's famous drawing "Vitruvian Man" literally embodies this ratio, which is also called the Golden Mean. That a pentacle is encoded in the English alphabet is evidence of illumined minds and esoteric society acting out an inter-generational plan and opening a special channel to

[58] Eisen, *The Universal Language of Cabalah*. pps. 77-80. See Appendix for details.

communicate across the centuries for those with the "eyes to see and ears to hear," who know where to look and what to listen for. During the Crusades, the Pentagram appeared on some of the coats of arms and shields of crusader knights. The pentacle encoded in the English Alphabet is a major clue to the Holy Grail from the time just prior to the origination of the Grail legends in the popular culture of Europe.

CHAPTER 12

Diotima's Beauty

The man [...] will suddenly have revealed to him
as he approaches the end of his initiation
a beauty whose nature is marvelous indeed,
the final goal, Socrates,
of all his previous efforts.
Diotima of Mantinea (Plato's Symposium)

No symbol has been more wrongly and persistently maligned than the one that the ancients associated with life itself; a symbol sacred to ancient people and many today, called the Blazing Star or pentagram. To the ancient Greeks the pentagram was a sacred symbol of light, health, and vitality. The pentagram was the badge worn by the Society of Pythagoras. The Greek author Iamblichus tells the story of a member of the Pythagorean Fellowship, who was far from home, and fell ill while staying at a wayside inn. Despite the care of the landlord, the traveler's illness was dire. Before his death, recognizing his desperate situation, and being unable to compensate his host, he asked his host for a board. He inscribed the pentagram on it and asked the host to place it where those who passed by would see it. Eventually, a Pythagorean brother was passing by on horseback and saw the symbol. He stopped, made inquiries, and upon hearing the story, he generously compensated the landlord.

The five-pointed star is a familiar geometric symbol today. Gold and silver stars are used in elementary schools to indicate achievement. It is used in the flags of some thirty-five countries. It appears on military vehicles, tanks, jets, and uniforms of all countries. The star is associated with law enforcement and is used in police and sheriff badges. It is also used as the symbol of the American Fraternal Order of Police. There are very good reasons why the pentagram is used by the police and the military. *The Divine Ration (Phi) ratio exists infinitely in the pentagram, and Phi exists in life.* Therefore the pentagram is a symbol of what the police and military protect - the lives of those whom they defend. During the late Medieval period finding a pentagram in someone's home would be adequate evidence of witchcraft and would result in torture and during the "Burning Times," the death penalty was by burning on the stake. By making the representatives of the law wear this image as their own symbol, such persecutions became much less likely. In "Dictionary of Symbols," (1994) by Karl G. Liungman he writes:

> From the sixteenth century until the eighteenth century the pentagon, as a structure closely associated with the pagan goddess through the celestial movements of the planet Venus, was common as a basic design for fortresses.[59]

Here is an incredible reason that for instance, the command center of the United States military is a

---

[59] Carl G. Liungman, *Dictionary of Symbols.* (New York, London: W.W. Norton & Company, 1991) p.314.

Pentagon complex in Arlington, Virginia. It is a reason that has to do with astronomy. Karl also wrote about the pentagram that, "The sign is the structure that results if one plots the movements of the planet Venus as seen from the Earth in the zodiac."[60] *This is not true as attempts to reproduce the pentagram this way will fail.* That failure is due to the error of the Earth centered perspective of the Cosmos and zodiac. This observation, that you must instead employ the heliocentric perspective to produce the pentagram, suggests that the Roman Church condemned the heliocentric model of the solar system as a heresy, punishable with death, *for this very specific reason.*

In his "Symposium," Plato records a dialog received from his teacher Socrates, of a dinner party attended by Socrates. The following quote is attributed to Socrates' instructress present at the party, a woman from Mantinea named Diotima.

> The man [...] who has directed his thoughts toward examples of beauty in due and orderly succession will suddenly have revealed to him *as he approaches the end of his initiation* a beauty whose nature is marvelous indeed, the final goal, Socrates, of all his previous efforts. This beauty is first of all external; it neither comes into being nor passes away; next, it is not beautiful in part and ugly in part, nor beautiful at one time and ugly at another. [...] He will see it as absolute, existing alone with itself, unique, external, and all other beautiful things as partaking of it. [...] This above all others, my dear Socrates, (the woman from Mantinea continued) is the region where a man's life should be spent, in the contemplation of absolute beauty.[61]

What Diotima called a "beauty" may begin with the Divine Ratio and continue with all that proceeds from it.

---

[60] Ibid, p.43.

[61] Plato, *Symposium*. Author's italics.

It is a mathematical beauty that a man may spend his entire life contemplating. One classic book is, "The Divine Proportion - A Study in Mathematical Beauty," by H. E. Huntley. The divine proportion or Golden Mean can be generated with greater and greater refinement through a simple mathematical progression called the Fibonacci sequence. The Fibonacci sequence begins with, 1, 1, 2, 3, 5, 8, 13, 21, 34. In this sequence, beginning with the third number each new number is the sum of the previous two. As the numbers increase in sequence, each two adjacent numbers create a ratio more closely approximate to Phi. About the Fibonacci sequence H.E. Huntley writes:

> A *pretty* result? What constitutes the essence of the aesthetic appeal of this outcome of simple mathematics? We seem to have no single word for it. It appears to be compounded of a mixture of archaic emotions. There is surprise at the unexpected encounter; there is also both curiosity and wonder -- making three of the flavors included in the idea of beauty. *Curiosity:* because one craves to understand why *Phi,* which permeates the pentagram and is at home in Platonic polyhedra, should also be the limit of a ratio initiated so casually and generated as described above, a series which is apparently not even remotely related to the geometry of the Greeks. *Wonder:* because the conviction grows stronger that we have chanced on an unexplored world which, like the universe around us, appears to have no boundaries. There must, we speculate, be other discoveries to be made here by the inquiring mind.[62]

The most extraordinary thing about the Phi ratio is that it abounds in nature and in the human body - and this is astonishing. According to Roger Jean, in "Phylotaxis: A Systematic Study in Plant Morphogenesis," "In a survey of nearly 13,000 observations of 650 species 96.5% are

---

[62] H.E. Huntley, *The Divine Proportion - A Study in Mathematical Beauty.* (New York: Dover Publications, Inc., 1970) p.36.

classic Fibonacci phylotaxis."[63] Phi reveals itself in nature in plant phylotaxis, which describes the arrangements of leaves on the plant. It is found in the florets in flowers such as the sunflower, pinecone, and pineapple. Mathematicians have studied this for hundreds of years. Phi's ubiquity in nature exceeds any chance of coincidence. Dividing the number of female bees by the number of male bees in a honeybee community will yield the Phi ratio. The Nautilus is a cephalopod mollusk that pumps gas into its chambered shell to adjust buoyancy. The ratio of each spiral shells diameter to the next is Phi. In the nautilus shell, the volume of each chamber is in Phi ratio to the next. In the daisy sunflower, each floret forms a dual duty in being part of both the right and left-handed spiral. This is remarkable enough, but it is astonishing that the numbers of florets in each spiral are adjoining Fibonacci numbers, twenty-one clockwise and thirty-four to the left. The dual role of the florets, the simplicity of the mathematical relationship and the amazing association with the Fibonacci sequence is mathematical beauty in nature. The Phi ratio is considered the most irrational number, because the difference between these fractions approach zero more slowly than any other irrational number such as Pi. There is no pattern to the resulting number sequence generated on to infinity. Thus, the Phi ratio most refuses to conform, which is a good reason for its association with the disobedient angel Lucifer, as it refuses to become graphed to a straight (orthodox) line. *Like life, this resistance to conformity is a reason why all life is self-similar, and yet not exactly alike.*

Leonardo DaVinci's drawing called the "Vitruvian Man" literally embodies the Phi ratio as it abounds in the human form. For instance, the navel divides our height into the

[63] Roger Jean, *Phylotaxis: A Systematic Study in Plant Morphogenesis.* (New York: Cambridge University Press 1994).

Golden Mean. The point of our fingertips to our wrist and the wrist to the elbow are in Phi ratio. Our shoulder to elbow and elbow to fingertips are Phi ratio. It is evident that Phi is one of the guiding principles of life. One of the mystical secrets encoded in the Vitruvian Man is the idea that man has a dual nature, one material and the other divine, or body and soul. This is shown in the proportions of man connecting to both the square, representing matter, and to the circle, representing divinity, perfection, and eternity. The square also represents earthly reason and the circle faith. In the Vitruvian Man, it is man himself that is the symbol of the pentagram. The Greek philosopher Protagoras of Abdera (490-420 BC) wrote, "Man is the measure of all things." If in man, the divine ratio exists and this same ratio exists in the heavens, then is the divine spark found in man? The train of thought leading to this began with the fact that some Norman cabal chose to create the English alphabet by adding the letters JUW in specific locations that resulted in encoding the Pi and Phi ratios in the alphabet itself and that this is an encoded pentacle. These things were the clues that lead to the secret of the Grail. After the Normans conquered England in 1066, they created the English alphabet and language by 1099. It is no coincidence the European obsession with the Grail immediately follows this period of history. It was by 1100-1118 that the Templar Order came into existence and the push to reclaim the Holy Land in the Crusades began. Now consider that the secret of the Grail is essentially the same thing as "Diotima's Beauty," and the "Royal Secret" of the 32nd and highest degree of Freemasonry.

Part 2 - So Below

In which the secret of the "name" of the Holy Grail as an astronomical phenomena read by Fleqetanis in the constellations is laid out plainly and how the same connection is made to Matthew 16:18 and other sacred symbolism, explaining why the Grail Secret implied by this "name" is the greatest heresy ever told.

CHAPTER 13

The Golden Compass

I imagine, that right now, you're feeling a bit like Alice, tumbling down the rabbit hole, hmm? [...] There is something wrong with the world. You don't know what it is, but it's there, like a splinter in your mind, driving you mad. [...] This is your last chance, after this there is no turning back. [...] Remember, all I am offering is the truth, nothing more.
Morpheus – The Matrix (1999)

There are two types of learning. One is by instruction, meaning *to build in*, to be told what to think and what to do. The second is by education,[64] which means etymologically *to lead out*, to direct us out of bondage into being free to think for ourselves. In the film "The Matrix" the character Morpheus, who is named for the Greek god of dreams, has the job of waking people up to the realization that their minds are not free. Morpheus offers a symbolic red or blue pill. The character Neo (neophyte) chooses the red pill (of initiation and esotericism) and learns to see things as they

[64] e, ex – out + ducere, duct - lead

really are, with eyes that hurt because they have never been used before. The drawback of being freed from the illusions of the world is the sobering reality that you may never be able to return into the illusion (of exotericism) and regain your innocence. The idea that we are imprisoned in a cell that we cannot see is a common philosophical idea that has existed at least since the time of Socrates who was executed for impiety. In "Plato's Republic," there is told the allegory of the cave. In this parable there are prisoners from birth who are chained by neck, hands, and feet in a cave facing away from the entrance. Shadows appear on the wall in front of them as their jailers project images of animals like a puppet show in front of a fire. This is all the prisoners have ever known of reality, and they see only by the dim light of a fire. One day, one of the prisoners is set free. He sees the world outside in bright sunlight, which hurts his eyes. Out of kindness he returns to the cave and tries to tell the others prisoners of things as they really are outside of the cave – but they treat him *as though he is insane*.

In "The Behavior of Crowds," Everett Dean Martin wrote, "there is a flaming hedge of logical abstractions, sanctions, taboos, which none but the intellectually courageous few dare cross."[65] If you cause people to question their basic beliefs, they may get angry or choose to think you are crazy to protect the way they had been taught to think. This is why there is a need for initiatory societies that prepare people for free thinking. Because for some, there is a "splinter in your mind, that drives you mad," for such it has been with the secret of the Grail; which has prodded many authors on; perhaps on the suspicion that it was a symbol of something forbidden; something that some people in power to this day, do not

---

[65] Everett Dean Martin, *The Behaviour of Crowds, A Psychological Study.* (New York and London: Harper and Brothers Publishers, 1920). p.156.

want the masses from ever knowing. Another metaphor is that people think inside a box (such as religious myth), and when you offer a hand to help them out of the box, they think you are crazy. But what some may consider crazy; others may recognize as an intellectual passion; perhaps bordering on obsession; to free people from the cave of their delusions. The Greeks have a word "aletheia," which has a number of meanings such as naked truth, unconcealed, unclosed (mind), and seeing things as they are. In 1995, the author Philip Pullman, published a novel called, "The Golden Compass." This novel is considered controversial because it employs allegories considered dangerous to simple faith. In the novel the main character has a fictional form of navigational compass of truth, called an "alethiometer." This is the compass of the title, but in reality there is such a thing as a "Golden Compass," also called a golden mean compass, with three legs, which automatically measure out the golden cut sections when the legs are moved in or out. The golden mean compass of the Phi ratio is one of the clues leading to this book's solution of the Grail secret. Perhaps this is the actual compass Philip Pullman refers to allegorically with his "alethiometer." You can use a golden mean compass and easily find examples of Phi in life, starting with your own arm. Today you can buy a golden mean compass in a toy store and use it to measure, for example, the distance from finger tip to wrist, and wrist to elbow. The Golden Section exists infinitely in the geometric figure of the pentagram star which many Christians would identify as a symbol of evil. The Golden Mean Compass is dangerous to faith because it contradicts this belief and reorients your attention away from evil and toward the positive direction of life itself.

The world has undergone a transformation in recent centuries. Some of the important changes have been those within the Roman Catholic Church. Change began back in the 16th century with the Reformation and the process initiated then continues today. In particular the Roman Catholic Church has turned around its earlier hostility toward science - and astronomy in particular. The Vatican currently supports and sponsors several modern astronomical observatories in research projects collectively called the Vatican Observatory. In 1981 the Roman Church founded the Vatican Observatory Research Group (VORG) in Tucson, Arizona in the United States. In collaboration with the Steward Observatory, the Vatican Observatory completed the construction of the Vatican Advanced Technological Telescope (VATT) on Mount Graham, Arizona, in 1993. This optical-infrared telescope, which has affectionately been dubbed, "The Pope Scope," pioneered new technology for creating large, lightweight, stable mirrors in a rotating furnace. The VATT is being used for long-term research programs. The Vatican's embracing of science and astronomy may be more than the pragmatic acknowledgment of scientific truth. Perhaps the claim since Vatican II, that religion cannot contradict science is a step toward acknowledging for the new millennium a radical new image on the faith. Who can say what the future holds, but it really is significant that the institution that once repudiated science is taking a role in its promulgation. The library at Castel Gandolfo possesses a valuable collection of rare antique books, including the works of Copernicus, Galileo, Newton, Kepler, Brahe, Clavius, and Secchi. The Vatican Observatory publishes its research results in international journals and distributes its annual report to more than 400 institutes around the world. With all this research, it would seem obvious the Vatican astronomers should know therefore, the Harmony of the spheres you are about to learn, *and will no longer reject this truth.*

The Phi ratio (1.6181) was enciphered in the English alphabet along with the ratio of Pi (3.1416) because it preserves a clue to this knowledge, which is a mystical secret of life.

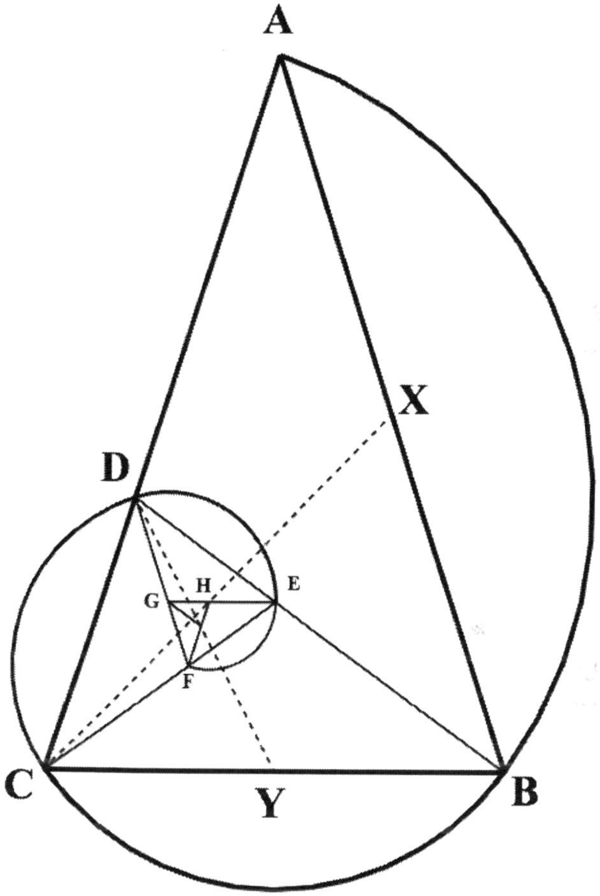

Figure 3 The Divine Triangle

As already explained the Phi ratio exists in many life forms on Earth, but it is not well known that the *Phi ratio also exists in the relation of the orbital periods of Earth and*

Venus. That Phi is associated with life on Earth and with the orbits of the heavenly spheres above seems to be a perfect expression of the Hermetic principle of Correspondence. "As above, so below; as below, so above." The divine triangle is called such because the two lines radiating from a single point that are exactly 36° apart will be in exact Phi ratio between the lengths of the lines and any equidistant endpoints. Given the divine triangle, if AB and AC are a length of 1.0, then BC is always 0.618 (called lesser phi with lowercase 'p'). If BC is length 1.0, then AB and AC are 1.618 (called greater Phi with capital 'P'). The apex angle is 36° and angles C and B are 72°. Bisecting the angle B produces another golden triangle as 72/2=36 in which the line bisects precisely at the golden cut at point D. If we rotate the new resulting triangle and keep bisecting the lower right angle, the golden triangle keeps reproducing itself infinitely in a logarithmic spiral around some central point. We can locate the point of convergence of the spiral, if we draw lines from the other base angles to the midpoint of the sides opposite to them (CX&DY). Thus the golden section appears infinitely in the divine triangle. This is important for understanding the symbolism of the Holy Grail as a source of food and life, as the divine triangle is the model for the ancient Greek concept of the *cornucopia*. On Earth, Phi produces the cornucopia of life. Notice that the logarithmic spiral, with each section increasing in size by the golden mean, is the same form in life as the Nautilus mollusk. Imagine, if you will, going back in time to the birth of the Nautilus spiral, in the diagram tracing counter-clockwise ABCDEFGH[...]. *We spiral down infinitely, to a source point that may philosophically be the origin of life.* This is like Alice falling down a rabbit hole where there are mathematical properties that are complex, but beautiful. In, "The Natural and the Supernatural," John Omar wrote,

"Beauty is a conspicuous element in the abstract completeness aimed at in the higher mathematics; it is the goal of physics as it seeks to construe the order of the universe; it ought at least to be the inspiration of all study of life [...]."

As the orbit of Venus is closer to the sun than the Earth's, it is never seen more than 48° from the sun. This means Venus can only be seen near the time of sunrise or sunset. For this reason, Venus is called both the Morning Star and the Evening Star. The ancients revered Venus as the only planet bright enough to cast a shadow. There are a number of names associated with Venus, including Astarte, Aphrodite, and Isis. It was also called the false light or the star of the morning or Lucifer, which means the light-bearer bringing forth the sun. Because of this relation to the sun, Venus is also called the Mother of the Gods. Venus is associated with the birth of the Son of God as she rises in the East and is the same Eastern Star or Bethlehem Star that guided the three wise men to the Christ child.[66] Interestingly, it is only at Christmas that Christ is associated with the star, which is usually depicted with 5 or 8 points and placed at the top of Christmas trees. The early observations of Venus through a telescope by Galileo showed that Venus goes through phases, as does our own moon. As Venus can even be seen in a full crescent phase, like our moon, it was proof of the Copernican Model, which placed the sun at the center of the solar system and not the Earth. A bit of thought on this makes it clear this could only exist in a heliocentric solar system. Only the outer planets can present fully illuminated spherical images especially when viewed at night when earthly viewers are facing away from the sun.

---

[66] There is a female branch of Freemasonry called the "Eastern Star" and their symbol is the inverted (downward pointing) pentagram. Christian tradition avoids association of the Bethlehem Star with Venus yet it could be argued this is purely a result of obscurantism leading to obfuscation.

An inner planet must always have some of its illuminated surface hidden from our view otherwise it is completely hidden behind the sun's glare. The harmony of the spheres is reflected in the life forms on Earth. The Golden Mean is ubiquitous in life forms on Earth, but much less known is that the Phi ratio exists in the orbital periods of Earth and Venus as discussed in the next chapter. This harmony and its mystical connection with life is indicated in the title of this book, "The Signpost of God" as it is an astronomically large indicator in the heavens above, which may lead humanity to a destiny in the stars.

CHAPTER 14

The Crown of Creation

Besides the lions on her gate, her symbol is an eight-pointed star.[67] In the Babylonian pantheon, she "was the divine personification of the planet Venus".[68]
Ishtar (wikipedia.org)

Pentagrams have been found on broken fragments of pottery in Palestine at strata dating from around 4000 BCE. The pentagram was common among ancient Sumerians around 2700 BCE. We also know with certainty, the pentagram was on the official seal of Jerusalem, and appeared as well over its main gate during the period of 300-150 BCE. The ancient Assyrians used an eight-pointed star as their symbol for Venus. Around 3200 BCE, Venus became associated with the five-pointed star. This indicates an advanced knowledge of astronomy, including a heliocentric cosmology. The astronomical

---

[67] Black, Jeremy and Green, Anthony (1992). *Gods, Demons, and Symbols of Ancient Mesopotamia: An Illustrated Dictionary*. ISBN 0-292-70794-0 pp. 156, 169-170.

[68] Guirand, Felix. "Assyro-Babylonian Mythology". *New Larousse Encyclopedia of Mythology* (trans. Aldington and Ames, London: Hamlyn, 1968), pp. 58, 49-72.

science behind these associations has been forbidden knowledge for most of the last two thousand years. Even in this day, professional astronomers may not be aware that simply tracing the planetary conjunctions of Earth and Venus produce the geometric figures exactly as depicted in this chapter.

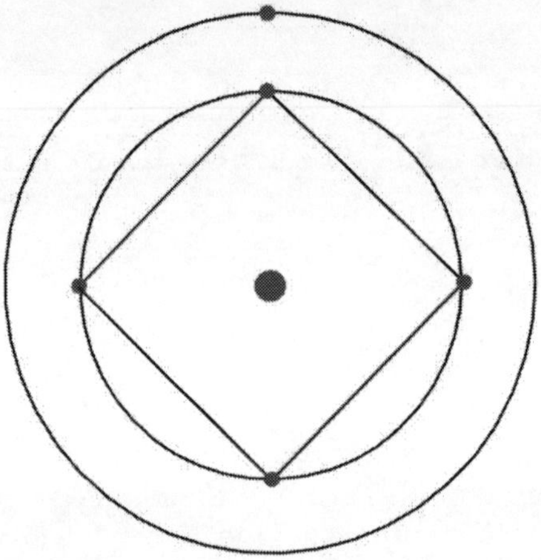

Figure 4 Venus - Earth 730-Day Harmonic[69]

In this chapter (and the Appendix) is the precise astronomical knowledge required for reproducing the astronomical tracings of Earth and Venus. The archaeological evidence linking Venus to these icons in the distant past are proof that the astronomical knowledge of forbidden Copernican (sun centered) cosmology was present at the dawn of civilization. It is this knowledge exactly, which has been most hidden or occulted to the masses of humanity. The most basic fact about Earth and Venus is that their two orbits are locked together in the

[69] The Venus orbital period is 224 Earth days. 730/224=>3.25 Venus years. Venus does 3 orbits + 0.25 or 1/4th orbit in precisely two Earth years. This places Venus (0.25x360°) +90° relative to its last conjunction with Earth after 730 Earth days.

Divine Ratio, the ratio of life on Earth. There are three types of astronomical conjunctions involving three or more heavenly objects. The astronomical term "superior conjunction" refers to the times when planets are exactly opposite each other with the sun between them.

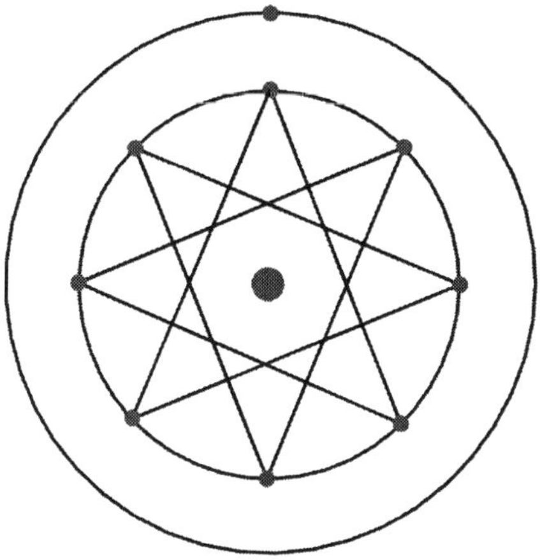

Figure 5 Venus - Earth 365-Day Harmonic[70]

The term "inferior conjunction" refers to those times when two planets are on the same side of the sun and in line with it. A "relative conjunction" describes a geometric relation, such as when two planets and the sun form a right angle (90°). All the examples provided here begin at the superior conjunction of Earth and Venus on June 6, 9:30 a.m. 2000 C.E. Eastern US time. An inferior conjunction could be used with similar results. You could use almost any modern astronomical program that displays planetary orbits, enter the date above, and advance the time in the

[70] The Venus orbital period is 224 Earth days and the Earth is 365 days. 365/224=>1.629 or 1 Venus orbit + ~226.6°. This is about 5/8th of an orbit counter clockwise or 3/8th orbit clockwise. After 8 years there are 8 points 45° apart. The Phi ratio exists in the 5/8th versus 3/8th relation splitting the Venus orbit in Golden Mean where 3 and 5 are Fibonacci numbers. This is how the Phi ratio exists in life forms such as the Octopus.

count of days indicated to trace the same harmonies shown graphically in this chapter. Such programs usually have a way to display the solar system rather than just the night sky, and that specific feature is needed for this purpose. The primary harmonic period between Earth and Venus is 2920 days or eight Earth years (8x365).

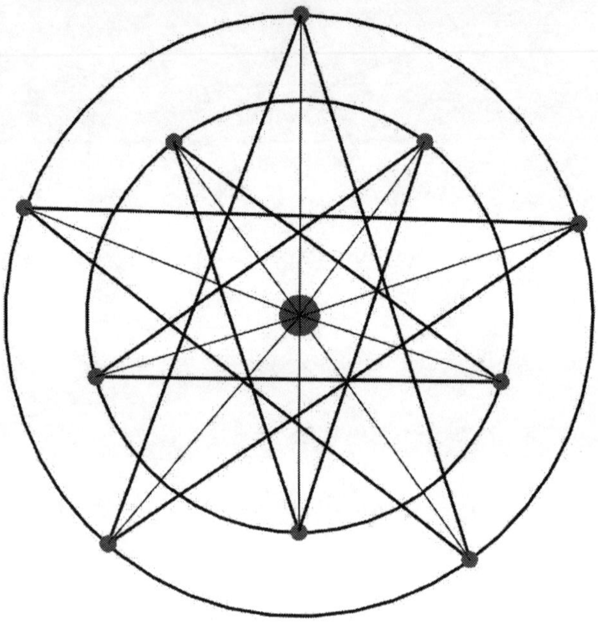

Figure 6 Venus - Earth 584-Day Harmonic[71]

[71] This astronomical period has a similar relation to both the Earth year and the Venus year for 584 days are precisely 1.6 Earth years (584 divided by 365) and 2.6 Venus years (584 divided by 224). This means that both Venus and Earth return to where they last where plus an additional 0.6 (phi) rotation. Since 0.6 times 360° equals 216° and 216° divided by 72° equals three, and as the number of degrees between each point of a pentagram is precisely 72°, the angle of each 584 day Earth and Venus conjunction traces the point of the pentagram star. It is possible this is a result of the Venus orbit being exactly 0.72 that of Earth's orbit and with the most perfect circular orbit.

Wherever Earth and Venus are in relation to each other, 2920 Earth days later, they will be in almost exactly the same place in their respective orbits. By starting in superior conjunction, Earth and Venus will be again in superior conjunction exactly eight Earth years later. *While Earth orbits the sun 8 times, Venus orbits the sun 13 times and 8 and 13 are Fibonacci numbers.*

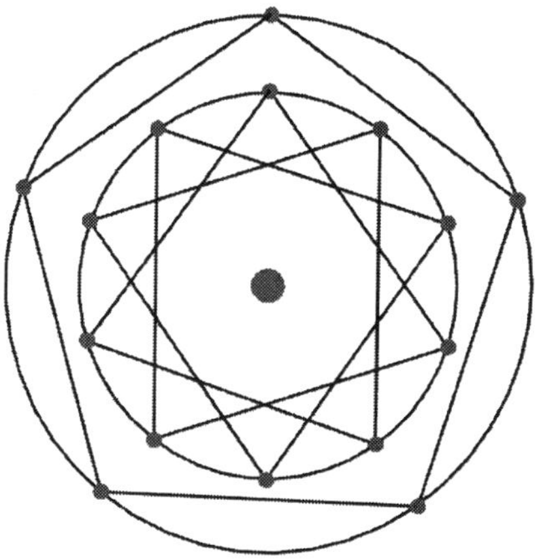

Figure 7 Venus - Earth 292-Day Harmonic[72]

The other harmonic periods are found in factoring 2920. The first factor is 2 and 2920 divided by 2 equals 1460, which is exactly four Earth years. If we start with a time when Venus is at its greatest height in the East as the Morning Star, then after this astronomical period Venus will be at its greatest height in the western sky as the

72 For Earth 292/365=>0.8, 0.8x360°=>288° degrees of orbit and 360°-288°=>72° short of a full circle which means for Earth the alternating superior and inferior conjunctions trace a pentagon clockwise. For Venus 292/224=>1.3 or 1 Venus year plus 0.3 orbit and 0.3x360°=108° or 3 times $1/10^{th}$ an orbit. As a result all conjunction positions of Venus are a factor of 36°, in order as 108°(3), 216°(6), 324°(9), 72°(2), 180°(5), 288°(8), 36°(1), 144°(4), 252°(7), 360°(10)°. Placed in the order of the common factor 36°, 72°, 108°, 144°, 188°, 216°, 252°, 288°, 324°, 360°.

Evening Star sometimes called Phosphoros or Hesperos. It is for this reason that the four-year period was special to the Greeks. They planned their Olympic games to coincide with these astronomical events. In this way the planet and goddess were thought to watch over the games. This is also the time period between which Venus goes from superior to inferior conjunction with Earth or vice versa. The second factor of 2920 is 4 and 2920 divided by 4 equals 730 days or two Earth years. By looking down from above the solar system, we see Earth as the outer ring and Venus as the inner. Up-and-down are relative terms in space, but for convenience, we are looking down on the solar system from the North. In this perspective, the planets orbit the Sun counter-clockwise and all the planets, except for Venus, spin counterclockwise. After every 730 days or two Earth years, Earth is back where it started and Venus is found 90 degrees counter-clockwise from its last position. This is an example of a relative conjunction. By drawing lines connecting each position, Venus draws a huge square in space after 2920 days. Another factor is 8 and 2920 divided by 8 equals 365 or one Earth year. Obviously, Earth is back in the same place every 365 days, but what Venus does is very interesting. By tracing the positions of Venus, we draw a precise eight-pointed star. The fifth Harmony is based upon the factor of five, as 2,920 divided by five equals 584 days. Both Venus and *also* Earth trace out a pentagram clockwise as you would by hand without lifting your pen from the paper. Note that if the chosen start date were an inferior conjunction of Earth and Venus, then the smaller pentagram of Venus would appear upright and exactly coincident with the larger Earth pentagram, rather than inverted as in the graphic. The opposed matching points on both pentagrams occur because of the planets being traced in superior conjunction with the sun. This was known as much as

6,000 years ago (4,000-3,200 BCE), proving this astronomical knowledge existed a very long time ago. You could never arrive at this knowledge unless you had a heliocentric model of the solar system. It is amazing this was known so long ago, yet only 400 years ago, this was forbidden knowledge punished with death.

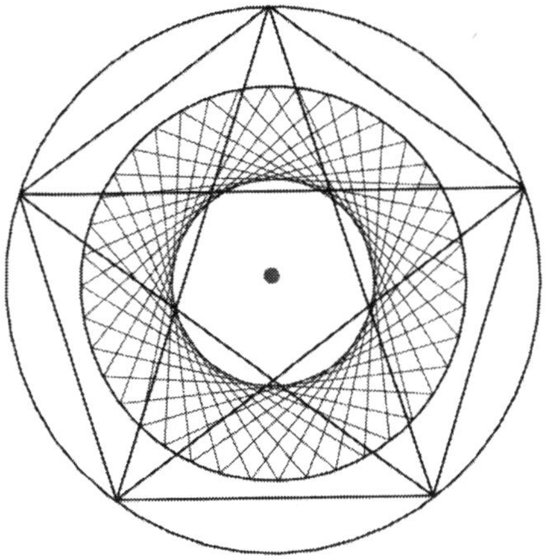

Figure 8 Venus - Earth 146, 73-Day Harmonics[73]

Venus and Earth are the only planets that can be unambiguously associated with precise graphic structures traced by plotting their movements in space. These orbital harmonics have been known since the time of the ancient Sumerians and Assyrians, as their symbols for Venus were at different times the same eight-pointed and pentagram stars. Another harmonic is based on the factor of ten, and 2,920 divided by ten equals 292 days. Starting with

---

[73] For Earth (pentagram) 146/365=>0.4, 0.4x360°=>144° (2x72°). So imagine starting with the top point and going counter clockwise 144° will trace to the lower right leg (as facing you), then to 288° to the left arm, then to 72° at the right arm, then 216° at the left leg, then 360° to the head. For Earth (pentagon) 73/365=>0.2, 0.2x360°=>72°. Therefore every 73 days Earth progresses exactly 1/5th of a circle (360°/72°=>5) or one side of a pentagon. For Venus 360°/40=>9°.

superior conjunction we trace Earth and Venus every 292 days. In this harmonic period Earth and Venus alternate between superior and inferior conjunction. During eight years Earth traces two pentagons (five sided regular polygons) clockwise, and Venus traces a ten-pointed figure clockwise.

Figure 9 Venus-Earth Rose[74]

The remaining obvious Earth-Venus harmonics are the 73 and 146-day harmonics.[75] The Earth traces a pentagram counterclockwise every two years using the 146 day harmonic or one twentieth of 2,920 days. The Earth orbit traces a counterclockwise Pentagon in one year using the 73 day period or one fortieth of 2,920 days. The 73 days divides evenly into all the other harmonics, 146 (two), 292 (four), 365 (five), 584 (eight), 730 (10), and 1460 (20), so therefore combining the harmonics of 146 and 73 days includes all the conjunctions. Tracing Venus in all 40

---

[74] By continuously plotting and tracing orbits of Earth and Venus together using a function which measures the distance between Earth and Venus over 8 years of time we see a graphical representation of their dance, which results in a geometric form commonly known as the Venus Rose.

[75] Research may show other harmonics related to the *rotation* of Venus or possibly Earth's *moon and menstrual* cycles.

points precisely 9° degrees apart, produces internally a circle, which very nearly meets the points of the inner Pentagon of the Pentacle traced by Earth. The computer-generated image is drawn to the exact spatial coordinates and the eccentricities of the orbits account for the slight offset of the inner circle.[76] If we compare the combined Venus and Earth 73 and 146 day harmonics with the symbol of the United States of America Fraternal Order of Police, it seems likely the designers of the symbol were aware of this astronomical harmonic. The circular band immediately inside the outer points, coincide with the orbit of Venus. The inner circle of the FOP Star (labeled "Jus-Fidus–Libertatum") corresponds to the circle traced by the 40 points of Venus. Relating the orbital harmonics with the Grail symbolism, recall that in the Robert de Boron story of the Grail it takes different forms and has a fivefold nature.

In the dance between Earth and Venus, two heavenly objects whirl eternally in proximity to each other, and together both generate a pentagram that may be thought of as a similar life form in greater and lesser size (like a boy and a girl), from two planets with the Phi ratio in their orbits. Johannes Kepler imagined Venus as female and Earth as male. In "The Harmonies of the World – Book Five," (1618) when speaking about the harmonies of the planetary orbits, Kepler identifies the Earth as male, "This difficulty is caused by the wedding of the Earth and Venus, or the male and the female. [...] Earth has kindly allowed her to ascend into aphelion or the Earth itself has descended into its perihelion toward Venus and as it were, into her embrace, for the sake of pleasure, and has laid aside for a while its shield and arms and all the works

[76] The software for this graphic was designed by the author. See the Appendices for the core library functions used to generate this data.

befitting a man; for at that time the consonance is minor."[77] The point should not be missed, that two heavenly *bodies* working in [orbital] *conjunction*, produce multiple unique, but similar offspring.

There is the greater Phi ratio 1.618, and its inverse 0.618 (called lesser phi with lowercase 'p'). The difference between them being '1' (Alpha-Male) and '0' (Omega-Female). The Phi/phi ratios are perfectly and complementarily inverse. With Earth over Venus 365/224 => ~1.618, and Venus over Earth 224/365 => ~0.618, this brings to mind "I am the Alpha and the Omega, the first and the last, the beginning and the end." (Rev 22:13)

This is also related to the Hermetic principle of Gender, "Gender is in everything; everything has its Masculine and Feminine Principles; Gender manifests on all planes."[78] May it be this knowledge that is the essence of what the mythologist Joseph Campbell imagined as the coordinating principle which was an, "*image of man's destiny as an organ participating in the organization of the Universe*," via an "*all-suffusing, all-informing principle or energy*?"[79] The truly amazing thing about the orbital conjunctions, is not that they trace regular figures, it is that they trace figures that are related to life on Earth. The essence of these geometric figures, and the essence of life on Earth is the common Phi ratio (1.618) that exists in the planetary orbits themselves and in the figures traced. The apparent cause and effect appears to evoke the Hermetic law of Correspondence, "As above, so below. As below, so above."[80] At the time these words are written, the

---

[77] Steven Hawking, *On the Shoulders of Giants*. (London, Philadelphia: Running Press, 2002) p.676.

[78] Kybalion, p.183.

[79] Campbell, *Flight of the Wild Gander*. p.154. (Author's italics.)

[80] Kybalion, p.113.

scientific search is on for finding an Earth-like planet. Is it possible that understanding the apparent connection between the planetary orbits and life below on Earth could be significant for detecting life elsewhere in the Cosmos? The tracings of the planets are measured in astronomical units that can be detected light years away. Is it possible that the planetary tracings could be the smoke that indicates the Blazing Star and fire of life exists in a distant planetary system?

CHAPTER 15

A Quantum Leap

If I have seen further,
it is by standing on the shoulders of giants.
Isaac Newton – 1676

About a century ago experiments were done that heated gases in a glass tube. The heated gases emitted light which was observed through glass prisms and had unexpected qualities. The light emitted was not like the broad rainbow that you may have seen in the sky after a storm or what you may have seen from a chandelier filled with prisms. The light displayed discrete vertical bars of color. These simple experiments led, not only to our colorful neon lights, but also to the greatest scientific discoveries of the twentieth century. For example, the physical laws discovered from these observations are now underpinning all the digital devices and telecommunications used in the world today. It also added to the science of Spectroscopy, whereby we can know the elemental atomic composition of distant stars and galaxies. The discrete vertical bars of light spectrum revealed the structure of matter on the tiniest scale.

Observing this the theoretical physicist Niels Bohr developed the foundations of our modern atomic theories known as Quantum Mechanics. In the early part of the last century, Bohr believed the pencil thin bars of light originated in the heart of matter itself, in the structure of the atom. He believed that atoms were like tiny solar systems, where the nucleus of the atom is like the sun and tinier particles called electrons were in orbit like the planets orbit the sun. However, whereas planets may orbit in any distance from the sun he posited that the electrons could only be in certain orbits, at fixed distances from the nucleus, and no others. Bohr suggested that when atoms were heated or cooled, electrons would jump from one fixed orbit to another, and when electrons went from an outer orbit to an inner orbit the atom would release energy in the form of a light photon, and this was why the atoms produced light only on specific narrow bands of color. This movement of the electron is not gradual. It is instantaneous. Unlike a real number with many fractions of a whole, it is digital in nature (like one whole finger at a time when counting). There is no in-between orbit and no transitional state of the electron. It is in one orbit in one instant and in another the next. It is from this phenomenon that we get the phrase "quantum leap." Bohr's theory was that, except for this quantum leap nature of the atom, it would instead produce an indistinct smear of color and not the narrow bands of color that experiments showed. There are very sharp reds and very sharp greens and the quantum leap originates the sharp color. The electron goes from orbit to orbit without any trace of being in transitional orbits. If you can imagine it, this would be like Earth's orbit instantaneously changing to the orbit of Mars. Bohr believed the nature of the atom was that its energy existed in discrete units that could not be sub-divided, which he called quanta, and this was why the electrons could only occupy discrete orbits. Quantum mechanics developed over the twentieth century and led to some very big inventions like lasers, transistors, integrated circuits and all electronics (named for the electron). Bohr developed his theory that *the number of electrons and their orbital positions around the nucleus determined the chemical properties and light emitted from the atom.*

Imagine the thought processes of Niels Bohr, which gave him the profound and provable insights that changed our world. His insight connected the narrow bands of light to quantum theories, which gave us modern computers and contributed to the ability to determine chemical and physical properties at interstellar distance. The evidence was there in the vertical bands of light in the spectrograph. This example suggests the possibility that there may be some future science, which may explain why the Phi ratio exists in the Earth and Venus orbits and is ubiquitous in life on Earth. Sometimes the recognition of anomalies that raise questions in the mind can lead to momentous scientific discoveries. Is there a real connection, which is science fiction now, and something that may be known in the next generation or a hundred years from now? By observing an atom emitting green and red light, we can determine the number of electrons and their orbital positions. Conversely then, if it were possible to directly observe the number of electrons and their orbital positions around the nucleus of an atom, we could predict the atom would produce green and red light - *without seeing the light through a prism.* Is it possible that the solar system, like the atom releasing a quantum of energy in the form of a light photon, also releases some form of energy yet undetected by science. Perhaps it broadcasts information, which may indicate “life” and “lifeless.” Imagine two possibilities. One is an unknown science similar to spectroscopy that would detect life at interstellar distance. The second is a more limiting possibility, that we may only detect life in our local neighborhood of stars by detecting the planets and their orbital properties around a sun. This second possibility is limited by the power of our telescopes. This is admittedly pure conjecture, yet so many things in the world today were science fiction an average lifetime ago.

Phi exists in the planetary orbits in this solar system and in the heavens above, and also in our bodies and life below on Earth. It seems like there must be some connection but what can it be? It so happens that within quantum mechanics there is an example of interconnectivity of matter without an explainable cause and effect. One of the strange predictions of quantum mechanics is something called "entanglement." The equations that emerged as a progression of quantum mathematics, predicted that if two "entangled" particles became separated by space they remain inextricably connected, with no limit placed on how far apart they become. One of the random properties of a particle is known as spin. In the case of entangled particles, if you measure one of the particles and found a left spin, then the other entangled particle will spin right. But the theory gets strange, because it predicts the very act of measuring the first particle will change the property of the second entangled particle *at the time of measurement*. According to theoretical physicist Brian Greene in "The Fabric of the Cosmos":

> While intuitively baffling, this phenomenon fully conforms to the laws of quantum mechanics, and was predicted using quantum mechanics long before the technology existed to do the experiment and observe, remarkably, that the prediction is correct. This sounds like voodoo; Einstein, who was among the first physicists to recognize – and sharply criticize – this possible feature of quantum mechanics, called it "spooky." [...] Something that happens

> over here can be entwined with something that happens over there even if nothing travels from here to there – and even if there isn't enough time for anything, even light, to travel between the events. [...] According to quantum theory and the many experiments that bear out its predictions, the quantum connection between two particles can persist even if they are on opposite sides of the Universe.[81]

In 1964 the Irish physicist John Bell devised a way to experimentally prove entanglement. Simply put, his insight was that if you randomly tested *different properties of the particles* (such as spin over various randomly chosen axis) over many iterations of experiment, that if the outcome was determined at the very time the particles became entangled, then you would get a positive outcome more than 50% of the time. If the outcome was truly random and occurred only at the time the measurement was made (as predicted by quantum mechanics), then it will be exactly 50% successful. The technology for testing this hypothesis did not exist in 1964, but it did in the 1970's and over the intervening decades has been proven true with increasing accuracy and distance. As told by Brian Greene:

> Pairs of appropriately prepared particles – they're called *entangled* particles – don't acquire their measured properties independently. They are like a pair of magical dice, one thrown in Atlantic City and the other in Las Vegas, each of which *randomly* comes up one number or another, yet the two of which somehow manage always to agree. Entangled particles act similarly, except they require no magic. *Entangled particles, even though spatially separate, do not operate autonomously.* [...] In 1997, Nicolas Gisin and his team at the University of Geneva carried out [an experiment...] in which the two detectors were placed 11 kilometers apart. The results were unchanged. On the

---

[81] Brian Greene, *The Fabric of the Cosmos, Space, Time, and the Texture of Reality*. (New York: Vintage Books 2004). p.80.

> microscopic scale of the photon's wavelengths, 11 kilometers is gargantuan. It might as well be 11 million kilometers – or 11 billion light-years, for that matter. There is every reason to believe that the correlation between the photons would persist no matter how far apart the detectors are placed.[82]

This is mind-boggling and seems to go against what we are taught by Einstein's "Special Theory of Relativity," by exceeding the speed limit of light. In High School we were taught that can not happen. But imagine if we are, on the atomic level, made of "star stuff" thrown out of a fiery stellar furnace, then maybe there are infinite entangled particles spread throughout the Cosmos, that can communicate across interstellar and galactic distance. To borrow a term from science fiction, these entangled particles may engender, "sub-space frequencies." The entangled particles aspect of quantum mechanics may be the key to communications unlimited by the speed of light. *The point here is that even if we do not detect an emanation connecting the Phi ratio in the planetary orbits with life on Earth in the near term, it does not mean none exists.*

If in Quantum theory the number of electrons and their orbital positions around the nucleus determined the chemical properties and light emitted from the atom, then is it possible the number of planets around a sun and their orbital positions may indicate the possibility of supporting life or not? Two hundred years ago, wireless communication and spectroscopy would have sounded like science fiction. Perhaps theoretical physics will someday provide an answer to how the orbital dynamics of Earth and Venus evolved, and why it is that the same Phi ratio existing in the planetary orbits exists in our own human bodies. Science is the study of cause and effect. Perhaps where there is smoke of life, there is fire, and an

---

[82] Ibid. p.114-115.

apparent cause and effect to detect life at inter-stellar distance. *The main reason the orbital harmonics will indicate life is the ubiquity of Phi/phi ratio in life and gender on earth, which may be as fundamental as carbon and water to life as we know it.* Like Neil Armstrong when stepping on the moon, might this potential discovery be conceptually, one small step for a human, and one quantum leap for humankind?

CHAPTER 16

A Star Called Wormwood

How you are fallen from heaven,
O Lucifer, son of the dawn!
How you are cut down to the ground,
you who weakened the nations!
Isa 14:12 KJV

Scripture says the birth of Jesus was foreshadowed by a celestial event. It is with the innocent Christ child that we associate the usual five or eight pointed "Star of Bethlehem" on Christmas trees and in nativity scenes. Certainly, the three wise men or magi revered the pentagram star as it is said the pentagram star is present at all initiations. The following gospel quote is the first association of Jesus to the bright Morning Star that rises in the east at dawn. "Now when Jesus was born in Bethlehem of Judaea in the days of Herod the king, behold, there came wise men from the east to Jerusalem. Saying, Where is he that is born King of the Jews? For we have seen *his star in the east*, and are come to worship him."

(Matt 2:1-2 KJV) When the wise men stand over the baby Jesus, it is as though the child and the star in the east are one and the same. "When they heard the king, they departed; and behold, the star which they had seen in the East went before them, till it came and stood over where the young Child was. When *they saw the star*, they rejoiced with exceedingly great joy." (Matt 2:9-10 NKJV)

In the Bible, the name "Morning Star" literally means the planet Venus. It is both the morning star in the east and the evening star in the west. It is possible that in certain latitudes and seasons someone with sharp eyesight can see Venus as a crescent. The crescent has been associated with the goddess since antiquity and the mythology of this does not associate with the phases of the moon. In her role as the "Evening Star," the goddess brings all things into balance. This is seen in the statue of "Blind Justice," a woman with eyes bound and holding a balance (evening) scale. The goddess has been associated with both the pentagram and the eight-pointed star due to the tracings of the planet Venus. In the Venus tablet of Ammisaduqa dated 1581 BC, Babylonian astronomers (Magi) refer to Venus as, "Nin-dar-an-na" or "bright queen of the sky" and recognized that both the morning and evening star were the same celestial object and supported this with detailed observations.[83]

Religiously conservative people think there is something bad about goddess worship, yet the largest practice of a form of Western goddess worship exists today in the Catholic cult of Mary. Ancient tradition describes the goddess having three phases of life as maiden, mother, and crone. The three aspects of the goddess are also represented in most cathedrals that have three front doors. The sanctuary space is designed to be

---

[83] Source wikipedia.org article on Venus.

womblike, and the usual three doors are elaborately carved stone openings that resemble female genitalia with a stone flower at the top representing the clitoris.[84] The larger central door represents the goddess in her ascendant phase as the mother of us all. If Christ had his way he would give the goddess to you, for it is written that he said, "And He that overcomes, and keeps my work till the end [...] I will give him the Morning Star. He that has an ear let him hear what the spirit says to the churches." (Rev 2:26-29) Reading between the lines we may see, "To you who overcomes [the prejudice to scientific truth] I [Jesus] will give [the forbidden teaching about the planet Venus also called] the Morning Star." The words, "and keeps my works till the end [...] I will give him the Morning Star" is in common with the pagan mystery schools. In Plato's Symposium Diotima of Mantinea remarks to Socrates that at the end of initiation the student will receive a "beauty" which is the final goal of all the student's effort and study. If the pagan traditions of initiation are equal to that offered by Christ, then what Diotima calls the "beauty" is equivalent to the "Morning Star" that Christ gives to you at the conclusion of the Bible. Therefore the Morning Star, the pentagram, the planet Venus, the goddess, the "beauty" of Diotima, and the Phi ratio are all implied as the same symbolically. In pagan religion, it was common for a god to have a female companion goddess. This created balance and equality between the sexes. The ascent of the one god concept arose out of strict patriarchal society and so God is imagined as a He and not a She. Even in Judea in the sixth century BC the Hebrews worshipped both a god and goddess. As monotheism and patriarchy advanced through time, references to this goddess became

[84] Church steeples are similarly phallic.

mythically changed into a symbol of evil as the religion shifted away from the polarity of male and female, to the dominant theme being good and evil.

According to Jacobus de Voragine's "Golden Legend," (1250 C.E.) in the south of France in Saint-Baume, there is a cave once inhabited by the Magdalene living as a hermit. She is supposed to have lived out her later years there naked, but clothed by her long hair. In Roman times, this same cave was the center for worship of the goddess Diana Lucifera called light bringer or Illuminatrix. The Roman name for the planet Venus as morning star is Lucifer and Vesper as the evening star. Other names for Lucifer are Astarte, Aphrodite, Isis, the false light, and the mother of the gods. Manly P. Hall wrote that, "Lucifer is the greatest mystery of symbolism."[85] There is a common misconception that Lucifer is Satan. Lucifer is not the Christian Satan, or evil. There is little to be found in the Bible about Lucifer. What is in the Old Testament about Lucifer is found in Isaiah Chapter 14. "How you are fallen from heaven, O Lucifer (or Morning Star), son of the dawn! How you are cut down to the ground, you who weakened the nations!" (Isa 14:12 KJV) There is no story of Angelic rebellion found here.

The third century BCE, translation of Isaiah, the Septuagint, translated the phrase "Morning Star" into Greek as Heosphoros. Almost two and a half thousand years after Isaiah, Jerome translated the Latin Vulgate with the Latin name Lucifer. Heosphoros, Lucifer, and Phosphoros as well, are all translations of "light-bearer" or Venus. If Lucifer was the goddess (Venus), and like Eve blamed for the fall of man and described as evil, is there some reason for this? The justification for this, such as it is, is found in the words, "you who weakened the nations."

---

[85] Hall, *Secret Teachings.* p.CXLV-ul

(Isa 14:12) Raising the goddess spiritually will ultimately have the effect of giving woman equality with men. Feminine values are considered weakening in opposition to male strength. In Genesis 3:16 God says to woman, "I will greatly multiply your pain in childbearing." In warrior cultures, woman are dominated and controlled for reproductive purpose and literally used for the raising of an army.[86] Over centuries of time Lucifer the morning star, morphed into a disobedient angel that was cast out of heaven to rule eternally in hell. In Christian mythology, Lucifer is now the same as Satan, the Christian devil, and ironically called the "Prince of Darkness." The descent of the shining angel from heaven to hell is mirrored in the descent of the goddess Venus into the underground of Western civilization. Lucifer is not a "he" but a "she" and called the most beautiful of angels – not Satan. Lucifer as a myth was originally a goddess, and evidence of this is the fact that Lucifer is associated with the pentagram or "Blazing Star" and this association comes from Venus and its orbital conjunctions with Earth. According to the scholarship of Elaine Pagels, Professor of Religion at Princeton University, the Book of Revelation was written, "around 90 C.E."[87] Therefore, as recorded in the book of Revelation, Venus and the pentagram star were cast out of heaven in the first century.

> And the third angel sounded, and there fell a great star from heaven, burning as it were a lamp, and it fell upon the third part of the rivers, and upon the fountains of waters; And the name of the star is called Wormwood: and the third part of the waters became wormwood; and many men died

---

[86] It is not a coincidence that "marital" and "martial" are anagrams (just reverse "it"). The root of both words, "mar," as in "Mars" (Roman god of war) means to "harm." When martial law takes effect, marital laws may change as well, particularly regarding birth control.

[87] Elaine Pagels, *Revelations, Vision, Prophecy, and Politics in the Book of Revelation*. (New York, NY USA Viking Penquin, 2012). p.47.

of the waters, because they were made bitter. Rev 8:10-11 KJV

The great star falls from heaven and it's burning as it falls is clearly a reference to the "Blazing Star." It is called Wormwood, and it makes bitter the water it falls on. Wormwood is a very bitter root from which comes absinthe. Absinthe is a mild hallucinogen that was reputed to be toxic to the brain. Absinthe is available in some alcoholic beverages to this day and may have contributed to the artist Vincent Van Gogh's madness and hallucinations, and yet inspired his paintings such as "The Starry Night" (1889). These attributes of wormwood, causing madness and hallucinations, clearly were the qualities associated with the mystical pentagram by the early Christians. The symbolism of this means the pentagram was deemed brain damaging, or rather faith damaging, to the common Adams and Eves. Indeed, during the Dark Ages the pentagram became associated with black magic, the devil or Satan, the beast of the Apocalypse, and the Antichrist. The pentagram would not be so associated with evil unless there was a powerful secret connected to it.

CHAPTER 17

The Mark of the Beast

And if thy right eye offend thee, pluck it out, and cast it from thee: for it is profitable for thee that one of thy members should perish, and not that thy whole body should be cast into hell. And if thy right hand offend thee, cut it off, and cast it from thee: for it is profitable for thee that one of thy members should perish, and not that thy whole body should be cast into hell.
Matthew 5:29-30 KJV

The Grail secret, as the astronomical knowledge that could be called the "Signpost of God," recalls the theme of the classic motion picture by Stanley Kubrick and Arthur C. Clarke, "2001: a Space Odyssey" (1968). At the opening of the film, the fanfare from Richard Strauss's "Also Spoke Zarathustra" accompanies a planetary conjunction of the moon, the earth, and the sun. The alignments of the planets represent the harmony of the spheres on the grandest scale imaginable. The movie begins with primitive ape-men. Overnight an advanced alien race places a black rectangular monumental slab in their camp with precise edges and smoothness. It is something completely out of their ken. The ape-men awake and become excited and shriek in confusion and amazement and tentatively run up and touch it. The leader of the ape-men soon after is inspired to use an animal's thighbone as a tool and weapon against other primitive ape-men. After killing the leader of a competing clan in a confrontation at a water hole, he throws the thighbone into the air where it tumbles in slow motion end over end. Then in one of the most dramatic moments in cinematic history, the bone becomes a long white space ship descending through black silent space. Kubrick neatly sums up four million years of human evolution with one splice of film. In the story, the same alien race has planted another similar monolith on the moon for humankind to find when their evolution and technology became advanced enough to find it there. When uncovered this monolith sends out a signal to the stars that contact has been made. Humanity is about to be led to their next level of evolution. Similarly, imagine the emotional reaction of the first astronomers who discovered and traced the planetary conjunctions of Earth and Venus some six thousand years ago. This is about the same time as the beginning of recorded history. It demonstrates real advanced knowledge of Copernican astronomy coincident with the advancement of writing. The Grail zeitgeist[88] emerged much later during the 12th and 13th centuries. This was the same time European society emerged out of the Dark Ages and into the Gothic Era in Europe. Coincidentally the building arts were then pushed to the limits of human engineering with massive cathedrals using flying buttresses that allowed enormous interior spaces to be lit by sunlight through stained glass windows. The Grail has since been hidden underground like the monolith buried on the moon in Arthur C. Clark's fiction.

---

[88] Zeitgeist – *German* noun. The spirit of the time; general characteristics of a particular period.

However there is a problem with science and religion. Science goes directly to the source of knowledge, bypassing any bureaucratic institution. Since it goes to the source, it was called Sorcery, which in medieval times was supposed to be a very bad thing. In fact, Christianity has strong anti-intellectual tendencies going back to the burning of the library of Alexandria. The word for this is "obscurantism," which means to be philosophically opposed to enlightenment. The word "obscurantism" comes from the Latin "obscurans," meaning a "darkening." It describes the practice of deliberately preventing the facts or full details of something being known. As theological power is based upon the Hermetic Principle of Correspondence, "As above, so below. As below, so above." It is the Theologians job to control how we perceive the heavens above; so astronomers have always been a problem to theologians. Astronomers project themselves by imagination into the heavens above. The astronomers seem to say in their heart, "I will ascend into the heavens [...] I will ascend above the heights of the clouds. I will be like the most High." (Isa 14:14) This is the sinful nature of Lucifer, and for this reason, many scientists and rationalists, who questioned and challenged authority in the name of a greater non-mythical Truth, were "brought down to Hell, to the depths of the pit," like Lucifer. (Isa 14:15) Men such as Giordano Bruno (~1548-1600) were killed for such things as claiming that the stars were all suns like our own and that there might be other solar systems like our own, harboring life as here. The Enlightenment scientist and astronomer Constantine Huygens (1596-1687) said, "The world is my country, science is my religion."[89] Scientists and Rationalists are autonomous thinkers, the very bane of conservative

---

[89] As related by Carl Sagan in the *Cosmos* documentary "Travellers Tales" copyright 2000.

religionists. Scientists always question and doubt the truths of their predecessors, the earlier "prophets" of science. This is antithetical to the mindset of religious people who may believe the Bible was literally dictated by God and cannot contain any errors or be questioned. Autonomous thinking rationalists and scientists are not easily influenced by mumbo-jumbo no matter how sophisticated the theology. There has historically been a large-scale conspiracy against the advancement of knowledge, and the myth of Lucifer is certainly influential in promoting fear of bringing enlightenment into the world. Perhaps if we can end all obscurantism we may stop setting limitations on knowledge and solve the world's problems. However there are obstacles placed in our path, and one of these is the connection between the symbolism of the myth of Lucifer, and another very scary myth called the "beast".

Even if you have not actually read Chapter 13 of Revelation, you are probably familiar with the "Beast of Revelation." Most know the number of the beast, 666 - and that it is supposed to represent the Antichrist. The number comes from Rev 13:18, "This calls for wisdom, let him who has understanding reckon the number of the beast, for it is a human number, its number is six hundred and sixty six." What mysterious meaning does the number 666 have? *The answer is it is an encoded representation of the Phi ratio.* The five points of the pentagram are each 36° and the angle itself is the angle of the Phi ratio because the distance from the point and the diverging two lines is always the Phi ratio. The connection of 666 to the pentagram is that if you take the primary angle of Phi or 36° degrees, and if you count the number, you get 666. Take all the numbers that you count from 1 to 36 (1+2+3+4[...]+36) and add them together and you will get 666 as the sum. This is the meaning of "count the number of the beast." (Rev 13:18) This is a truly remarkable association between the Phi ratio and 666. By geometrically translating the most irrational of numbers, which continues infinitely and never repeats a number sequence, we arrive at a very rational whole number that does that exactly. It is a human number because 666 as a symbol of Phi is throughout ones own human form. Also consider, ***"the beast was given a mouth uttering haughty and blasphemous words, and it was allowed to exercise authority for forty-two months."*** (Rev 13:5 KJV) It may be argued this cryptically refers to Jesus who taught the grail knowledge and also blasphemed, as forty-two months or 3.5 years is the commonly supposed length of time attributed to his earthly ministry.

$$\sum_{i=1}^{36} i = \frac{n(n+1)}{2} = (36 \times 37) / 2 = 666$$

Figure 10 Formula for 36° = 666 = 1.618

The gospel of Mark also contains extremely interesting clues regarding the "mark of the beast". (Rev 13:16) A careful counting of the number of verses in the King James Bible shows the number of the verses adding up to exactly 666 verses. There is a difference with some versions of Mark in which this is known as the shorter version. The difference is in the last chapter only. That chapter is 16 which has 8 versus rather than others that have 20 verses and the shorter version is considered the more reliable. This may be a clue of 16-8 of which relates to Phi 1.618 *or another verse in the gospels.* The longer versions of the gospel of Mark are considered later historically and contain a very significant difference, which strongly supports the Grail thesis of this book. Because the longer versions contain a verse 16:18 and that verse contains, "they will pick up serpents, and if they drink any deadly thing, it will not hurt them; they will lay hands on the sick, and they will recover". This indicates a teaching which may be forbidden and the poison is a further association with Jesus as the serpent containing some venom, which is also like a cure to a disease or delusion, like the star cast out of heaven and called wormwood. The Coptic Orthodox Church has its roots in Alexandria Egypt; as the tradition is Mark the Evangelist established the church; and Mark was its first bishop in 49 AD. There are some interesting traditions that would be fruitful for research. Mark himself, is reputed to have been born in the Greek city of Cyrene, one of the Greek Pentapolis settlements of northern Africa. Coptic tradition also holds that Mark the Evangelist hosted the disciples in his house after Jesus' death; that the resurrected Jesus came to Mark's house (John 20); and that the Holy Spirit descended on the disciples at Pentecost in the same house. There are also interesting implications regarding St. Catherine and St. Catherine's monastery at the base of the mountain reputed to be the one upon which Moses received the Ten Commandments; as St. Catherine is widely regarded as a fabrication as a counter to the female pagan philosopher Hypatia of Alexandria; a teacher at the Library of Alexandria who was by brutally murdered by the early Christians. One account of St. Catherine says, "According to tradition, angels took her remains to Mount Sinai. Around the year 800, monks from the Sinai Monastery found her remains."[90] This is one of the oldest Christian monasteries; it contains the oldest continually existing library in the world, and possibly this myth indicates some

---

[90]Wikipedia.org on Saint Catherine's Monastery, 8/26/2017

special preserved knowledge about Hypatia of Alexandria, or something preserved from the burned Library of Alexandria. The association of Mark the Evangelist with the mark of the beast and 666 strongly implies Mark had some significant knowledge; the extended version of Mark which in verse 16:18 mentions a poison which can be drunk with immunity; implies this is forbidden knowledge; which has special power to heal illness. This is a clue to the Grail knowledge.

It is amazing that the most irrational and infinite Phi ratio emanates from an angle (36° degrees) that is a simple whole number. Thirty-six degrees is also encoded in the celebration of the Feast of the Epiphany where *three* magi visit the Christ child in a manger on the *sixth* of January and 3 joined to 6 is 36. Remember that 36° degrees is the primary angle of the pentagram, and of the three magi in the gospels it is written, "When they saw the [pentagram] star, they rejoiced with exceeding great joy." (Matt 2:10) *Might this be the very epiphany of the eponymous holiday, which connects this holiday with pagan traditions?* According to Albert Pike, “The star which guided them is that same Blazing Star, the image whereof we find in all initiations [...] the Sacred Pentagram. [...] When Truth comes into the world, the Star of Knowledge advises the Magi of it, and they hasten to adore the Infant who creates the future.”[91] Thus, 36° degrees and the Phi ratio are associated with the Christ child and with Venus since the Magi followed the Eastern Star. This is the same as the Morning Star that is the planet Venus/Lucifer as it appears in the morning presaging the day or a time of enlightenment. The 36° degree angle is the origination point where life enters the universe and is implied as the Gnostic Pleroma.

---

[91] Pike, *Morals and Dogma*. p.842.

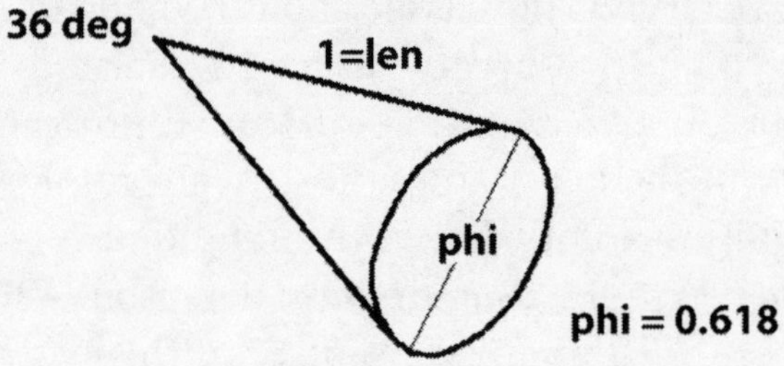

From the Grail comes life on Earth. Wolfram von Eschenbach, "I mean whatever the Earth yields." 666 = 1+2+...36, 36 degrees => Phi Ratio is the Cornucopia of Abundance, the Amalthean Horn.

Figure 11 Phi Ratio as Cornucopia

The point of the pentagram star is likely the source of the myth of the Amalthean horn of plenty. It is also known as the Harvest Cone or cornucopia and has been a symbol of abundance dating back to the 5th century BCE in Greece. Unlike anything else, this is the perfect interpretation of the Grail symbolism. In all stories, the Grail is supposed to provide food and drink of Paradisal excellence. According to Wolfram von Eschenbach, "I mean whatever the earth yields."[92] The Phi ratio simply fits this symbolism better than any other idea. Unfortunately, it is also associated with the Beast of Revelation and 666.

The denial and fear of the pentagram traced by Earth and Venus conjunctions may have led to the persecution of astronomy. Historically, we are faced with the dilemma of well meaning but self-righteous people, who are offended by truth, and who attack anything that penetrates their mythical delusions. In this chapter's quote of Matthew 5:29-30, we read about how it may be profitable to pluck out our right eye or cut off our right hand to save our lives. As in extremism and politics,

[92] Parzival, p.240

sometimes the right is wrong and the left correct and vice-versa. To the detriment of western civilization, theologians outlawed heliocentrism, the pagan mysteries, and truth about the commonality of Christianity with earlier pagan faith. Because of this denial of truth – there were book burnings *and the Dark Ages were the result.* The pagan Magi, were initiates and astronomers who understood the importance of the Blazing Star. The star of Venus helped to lift humanity out of Neolithic culture when humans finally understood the planetary orbits and our place in the Cosmos. May the Grail zeitgeist emerge again in common conscience to inspire humanity to the next stage of evolution and a destiny in the stars? As Albert Pike indicated, both faith and reason rest upon the Holy Scripture and combine to form the Blazing Star of Truth, *and now we proceed to show how this is true.*

CHAPTER 18

The Keystone Secret

The stone which the builders rejected
is become the head stone of the corner.
This is the LORD's doing;
it is marvelous in our eyes.
Ps 118:22-23 KJV

The keystone is the final stone put in place when building a stone archway, and the secret of its making has always been part of the knowledge of stonemasons since great antiquity. In fact, it was ancient tradition that the master mason would stand beneath the archway when the scaffolding was removed to suffer the consequence if it were not constructed to properly distribute the load. In "The Hidden Church of the Holy Graal," A.E. Waite writes about Parzival's stone, "This stone is the Key of the Royal Arch."[93] In this chapter is the keystone argument upon which this Grail thesis will stand the test of time. One obvious place to look for esoteric understanding of the meaning of the keystone is in "Morals and Dogma," by Albert Pike in the thirteenth

[93] Waite, *The Hidden Church*. p.395.

degree, called the "Royal Arch of Solomon." In this degree, it is plainly encouraged that the reader should make their own conclusions about the esoteric interpretations of symbolism, and that the keystone of civilization is something sacred, which cannot be hastily changed or violated with impunity, such as a constitution which preserves liberty.

As shown earlier, the English alphabet has the mathematical ratios of Pi and Phi encoded in the symmetrical and asymmetrical characters. Pi is the ratio of that symbol of perfection - the circle. The circle is associated with Infinity and eternal self-regeneration in the symbol of the serpent arranged in a circle and consuming its own tail. The circle and Pi have been associated with God in the circular statement "I am that I am." This is found in Exodus 3:14, where 3.14 would have been about as accurate a value for Pi as may have been achieved in Old Testament times. In the Gospel of Matthew, Christ says Peter is the rock on which he will build his church. He says, "the gates of hell," and in some translations "the powers of death," shall not prevail against it.

This is the verse traditionally used by the Catholic Church to claim special authority based upon apostolic succession. Does it make sense to interpret the rock as a mortal person (apostle Peter) or rather a principal such as the eternal regeneration of life? The Phi ratio is clearly is an expression of some divine principal underlying life, and representative of eternal regeneration and the evolution of life. By consideration of religious significance, any symbol of regeneration is sacred and a verse declaring the foundation stone of the church must be significant. It is somewhat commonly recognized that the circular logic of, "I am that I am," is not by accident associated with the value of Pi, as the verse is Exodus 3:14 or 3.14, which is Pi.

The keystone is a rock, and Wolfram von Eschenbach's grail is a rock. *Knowing the importance of regeneration and its connection to the Phi ratio can it be ignored that the verse upon which the church is built is Matthew 16:18 or 1.618, which is the Phi ratio?* "And I tell you, you are Peter, and on this rock I will build my church, and the powers of death shall not prevail against it." (Matt 16:18) As you continue to read ask yourself, "Is there is any way that this could be a coincidence?" Is this the secret foundation stone of the Christian teaching denied all these centuries because it is so closely associated with the ancient pagan mystery teachings and the pentagram star? For centuries, the Catholic Church has made claims of superiority to other Churches based on a literal interpretation of this verse from Matthew and apostolic succession from a human Peter. Any other interpretation has been heretical until now. Associating the Phi ratio as the rock of the Church is very close to declaring the pentacle or pentagram as the rock. The Phi ratio (1.618) as the interpretation of the rock of Matthew 16:18 upon which Christ builds his church strongly implies the connection of Jesus with the pagan Mysteries and the esoteric pagan schools persecuted and destroyed by the early Pauline Christians. The very name "Peter" or "Petros" translates as "rock" and also has common phallic slang associations and the Grail as rock being 1.618 also points to regeneration and implies higher mathematics and the pre-Christian mystery school teachings. According to Manly P. Hall:

> If it is ever established beyond question that Jesus was an initiate of the pagan Greek or Asiatic Mysteries, the effect upon the more conservative members of the Christian faith is likely to be cataclysmic. If Jesus was God incarnate, as the solemn councils of the church discovered, why is he referred to in the New Testament as "called of God an high priest after the order of Melchizedec"? [Heb 5:10] The words "after the order" make Jesus one of a line or order of

> which there must have been others of equal or even superior dignity.[94]

In that other theory of the Grail being about Mary Magdalene, it is commonly proposed that esoteric societies existed to protect the descendants or bloodline of Jesus. This is not the most reasonable explanation. If men like Leonardo DaVinci were grand masters of a secret order then their charge was not to preserve a "bloodline" in a genealogical descent, but rather to preserve the sacred knowledge that is symbolized by blood. In this way, esoteric societies are not about Jesus (as some might claim), but rather Jesus himself was likely a member of an esoteric order the likes of which have always existed since the dawn of time. Esoteric schools are like the Kingdom of Heaven in that they are all around but unseen, and of which you must pass tests to gain admission or initiation. Entrance into an initiatory order of advanced learning is for only the few, "Because narrow is the gate and difficult is the way which leads to life, and there are few who find it."[95] Here "life" is a codeword for esotericism. The implication then, is Christ was a master in the same sense as an esoteric instructor of the mysteries. Then an unseen hand representing an esoteric order (such as the Orphics) placed Matt 16:18 in its precise position, and that this is the ancient and essential order followed throughout history, whereby the advanced minds of the distant past communicate with the present day.

If the Phi ratio is part of the foundation of the church then modern Christianity has become significantly disconnected from its roots, because the perfect symbol of the Phi ratio is the pentagram, which Christians have historically associated with evil. The English alphabet was

---

[94] Hall, *Secret Teachings*. p.CLXXVIII.

[95] Matt 7:14 NJKV. See Ouspensky, "A New Model of the Universe," chapter "Christianity and the New Testament."

designed in the generation after the Norman Conquest in 1066. When thc Pi and Phi ratios were encoded in the English alphabet, what they really encoded was the circle and pentagram which when combined becomes the pentacle. We know, "As above so below. As below so above," that the Phi ratio exists in the orbits of Earth and Venus, and generates geometric shapes that also exist in life below on earth. We can reason therefore, that there must be some mystical connection between these things. What is today a mystical connection may tomorrow be an understood science of an emanation between the orbital harmonics and life on earth. Perhaps if we can fully understand why the same mathematical ratios and forms exist in the Earth/Venus harmonics and *also* life on earth, then we may be able to detect life forms at a great distance - something that is today only found in science fiction. This knowledge is suitable as a cup of Christ, containing his blood, as blood is symbolic of knowledge. Moreover, this knowledge could clearly have been the secret teaching of Christ as a scientist connecting him to the Neo-Platonic schools such as the Alexandrian "Museum" of his day and the Greek traditions. That Wolfram von Eschenbach described the Grail as a rock strongly suggests the verse Matthew 16:18 (1.618 Phi) refers to the rock upon which Christ built his church. The Grail as described here, as the Signpost of God is a perfect explanation of the psychological and historical enmity toward science and astronomy in the Dark Ages. That Matthew 16:18 is a reference to the rock being the Phi ratio upon which the church was built, has previously been a challenge to the authority of the Catholic Church and also the keystone of the argument that the secret of the Grail is *the greatest heresy ever told.*

CHAPTER 19

The Morphing Orphics

Through numerical coding of key words and phrases in the Greek (New) Testament, Christianity was linked with the Pythagorean tradition, in which music and number were the first principles of the universe.
Joscelyn Godwin Ph.D.

While the early Christians were trying to Christianize the pagan world, the pagan initiates responded with a powerful intellectual effort to Paganize Christianity. While the former largely succeeded in promoting a Christian identity, the latter succeeded as well. Pagan symbolism and doctrines are irrevocably intertwined with Christian religion. When it became unhealthy to be openly pagan, the initiated hierophants of Pagandom simply transferred all their teaching into the newly named cult. The word "revelation" depends etymologically on a "re-veiling" which means to put on a new garment or hide behind a new veil. This means the hierophants merely assumed a new Christian appearance or *put on the costume* of the new religious mythology. In his book, "The Golden Thread," the author Joscelyn Godwin identifies the Orphics as the pagan initiates who re-veiled the ancient teachings.[96] This was an important Greek esoteric tradition based on the philosophy of Orpheus. The Orphic philosophers had very Christian-like aspects. For one thing, they proselytized their mysteries. They transmitted some knowledge that was directly helpful in facing the prospect of death. They also taught that our actions in life would affect our hereafter existence. The Orphics believed in a gentle persuasion to make humans better and more civilized rather than using a forceful psychology or threats. A human Orpheus living before the time of the Trojan War, attempted to reform the Dionysian tradition in much the same way that Jesus tried to reform the Mosaic tradition. In the early Christian era, the Orphics experienced a revival in their influence. According to Godwin:

> The Orphics had been the first philosophers of Greece and the spiritual ancestors of the Pythagorean and Platonic schools, renowned for asceticism and for belief in the immortality of the soul. Now, in the Orphic revival, they stamped their principles on the new religion. *Through numerical coding of key words and phrases in the Greek (New) Testament, Christianity was linked with the Pythagorean tradition, in which music and number were the first principles of the universe.* [author's italics][97]

Note that phrases in the New Testament were numerically encoded to link Christianity to the

---

[96] For supporting evidence of this Godwin noted the works of William Sterling, "*The Canon: An Exposition of the Pagan Mystery Perpetuated in the Cabala as the Rule of All the Arts*" (first ed., 1897, reprinted London: Garnstone Press, 1974; and John Mitchell, "*The Dimensions of Paradise: The Proportions and Symbolic Numbers of Ancient Cosmology*," (London: Thames & Hudson, 1988); and David Fideler, "*Jesus Christ, Sun of God.*" (Wheaton, IL: Quest, 1993).

[97] Joscelyn Godwin, *The Golden Thread, The Ageless Wisdom of the Western Mystery Traditions.* (Wheaton IL: Quest Books – Theosophical Publishing House, 2007).

Pythagorean tradition. For example, Christ or a savior being born on December 25th is possibly the most commonly recognizable feature of the ancient pagan faiths. Even the New Advent Catholic Encyclopedia admits that when Christmas began to be celebrated in the fourth century, they did so because the date was sacred to the heathen cults. According to Kersey Graves in, "The World's Sixteen Crucified Saviors or Christianity Before Christ":

> Bacchus of Egypt, Bacchus of Greece, Adonis of Greece, Krishna of India, Chang-ti of China, Chris of Chaldea, Mithra of Persia, Sakia of India, Jao Wapaul (a crucified Savior of ancient Britain), were all born on the twenty-fifth of December, according to their respective histories.[98]

When the time came to assign a date to celebrate the birth of Jesus, December twenty-fifth was chosen because of very ancient tradition associated with the Gnostic Pleroma. Albert Pike wrote, "It is not in the books of the philosophers, but in the religious symbolism of the Ancients, that we must look for the footprints of Science, and rediscover the Mysteries of Knowledge."[99] For one thing, December is the month of the winter solstice in the northern hemisphere. The sun dies on the winter solstice December 21st and the waxing is not visible to the naked eye for three days and is re-born on December 25th. Thus, it represents the rebirth of the cycle of life in a bioregional mythology focused on the northern hemisphere. The year is the orbit of the earth around the sun, which is the circle of the year through the zodiac. December 25th is the 359th day marking the end of the circle of 360 degrees. The remaining intercalary days are the time of feasting and festival of the Pleroma. Also, the twelve days of Christmas

---

[98] Graves, *Sixteen Crucified Saviors.* p.69.

[99] Pike, *Morals and Dogma.* p.842

ending on January 6th, represent the twelve zodiacs. The 3 wise men visiting the baby Jesus on the celebration of the Epiphany on January 6th is the 36° degrees of the pentacle star, and of the three magi in gospel, "when they saw the star, they rejoiced with exceedingly great joy." (Matt 2:10) The 25th of December is 5x5 and both Earth and Venus trace the pentagram star in the heavens. The zodiacal symbol of December twenty-fifth is Capricorn, the goat. The goat, like Christ, is also symbolic of sin, "We know this man [Jesus] is a sinner." (John 9:24) The goat was sent into the wilderness to die as a sacrifice or scapegoat for the sins of the people and Christ is said to have died on the cross as a scapegoat for our sins.

A principle put forth in the Royal Arch of Solomon Degree is, "Whatsoever a man soweth, that also he shall reap." The degree makes plain that there are great esoteric truths concealed in legend, myth, allegories, and symbolism. These hidden truths have existed since the earliest time of civilization; yet are unknown to the world at large. In the text it speaks about what happens when we are tempted to do the wrong thing and a seed of evil is planted. To wit, "To-morrow it will threaten discovery. Agitated and alarmed, we cover the sin, and bury it deep in falsehood and hypocrisy. In the bosom where it lies concealed, in the fertile soil of kindred vices, that sin dies not, but thrives and grows; and other and still other germs of evil gather around the accursed root; until, from that single seed of corruption, there springs up in the soul all that is horrible in habitual lying, knavery, or vice. [...] The evil deed may be done in a single moment; but conscience never dies, memory never sleeps; guilt never can become innocence; and remorse can never whisper peace."[100] The denial of the secret of the Grail, or what in later history

---

[100] Pike, *Morals and Dogma*. p.217

became symbolized by the name "Holy Grail" to preserve it, was denied by about 90AD. This is when modern scholarship dates the authoring of the book of "Revelation."[101] It records a "burning star" being cast out of heaven and resulting in bitterness and death for many people.[102] This was the seed of evil planted, and once set in stone (so to speak) or dogma, the accursed root of evil drew around it and nurtured all sorts of evils. If Jesus taught science, the Grail knowledge shows how antithetical Christianity became to the original teachings of the master.

Surprisingly, the miracles regarding the fish and loaves of bread can be interpreted as astronomical teaching, especially because when afterward the disciples missing the point ask him about it and he says, "Having eyes, do you not see? And having ears, do you not hear? And do you not remember? When I broke the five loaves for the five thousand, how many baskets full of fragments did you take up?" They said to Him, 'Twelve.' Also, when I broke the seven for the four thousand, how many large baskets full of fragments did you take up? And they said, 'Seven.' So He said to them, How is it you do not understand?"[103] The keys to understand this are the numbers answered being 12 and 7. These two numbers indicate that the miracles implied, that He was really teaching about the stars and the planets, or the Universe itself. There are 12 zodiacal symbols and 7 planets in the astronomy of the ancient world. In the astronomy of the classical period of history, the planets were the objects in the sky that moved against the fixed background of the stars. They did not include the Earth as a planet but did include the Sun and the Earth's moon as planetary bodies. Therefore, the

[101] Pagels, *Revelations*, p.47.

[102] Rev 8:10-11.

[103] Mark 8:18-21 NKJV

planets included the Sun, Moon, Mercury, Venus, Mars, Jupiter, and Saturn. In the Roman Empire these were also associated with the seven days of the week and symbolically also with the seven metals of the classic period and the polytheistic names of the Roman gods as well. When the Roman Empire expanded they interpreted the Scandinavian and Germanic god names with like characteristics with the Roman gods. This practice was called the "Interpretatio Romana" (or Germanica depending on your perspective). Since the English language is an Anglo-Saxon language and the Saxons were a German people, the English weekday names are based in part upon the ancient German god names. Therefore; Sunday is Sun Day or Sunni's Day and associated with gold, Monday is Moon Day or Mona's Day and associated with silver, Tuesday is Tyr's Day or Roman Mars Day and associated with iron, Wednesday is Wodan or Odin's Day or Roman Mercury's Day and associated with quicksilver, Thursday is Thor's Day or Roman Jupiter's Day and associated with tin, Friday is Frigga or Frige's Day, or Roman Lucifer and Greek Venus Day associated with copper, and Saturday is Roman Saturn's Day associated with lead.

The denial of science along with the Grail, which became buried in the occult underground, became as a seed of evil, which may have caused the destruction of Truth and the very downfall of the Roman Empire, resulting in the Dark Ages and ultimately leading to the Burning Times. The evils that resulted included the destruction and censorship of knowledge, burning people on the stake for witchcraft, the persecution of intellectualism or the free pursuit of knowledge, and particularly making helio-centrism a heresy punishable by death. To fix this problem, the evil needs to be uprooted and brought into the light, and to do so we must find that which has been lost and restore it. The denial of knowledge of the Grail is the seed of evil. But imagine if we can take the stone the builders of the church rejected and make it the keystone of the corner, then perhaps we would say, "it is marvelous in our eyes."[104] Perhaps mysticism will lead the way as the middle path between science and spirituality, and resolve the ancient conflicts between science and religion, so that we may resolve in ourselves our faith and reason. Then as a result, we can honor our classical pagan inheritance.

---

[104] Ps 118:23.

CHAPTER 20

Great Instauration of Light

You shall not make for yourself a graven image,
or any likeness of anything that is in heaven above,
or that is in the earth beneath,
or that is in the water under the earth;
you shall not bow down to them or serve them.
Exodus 20:4-5 RSV

One day while teaching a high school class, Johannes Kepler had what he believed to be a divine revelation of God's hidden geometric construction of the solar system. This occurred on July 19, 1595 when the thought occurred to him that the orbital spheres of the planets, Mercury, Venus, Earth, Mars, Jupiter, and Saturn, could be described by the five Platonic solids. Kepler published this theory in his book with the title, "Mysterium Cosmographicum" or "The Sacred Mystery of the Cosmos" in 1597. He wrote:

> The Earth is the measure for all other orbits. Circum-scribe a twelve-sided regular solid [dodecahedron] about it; the

> sphere stretched around this will be that of Mars. Let the orbit of Mars be circumscribed by a four-sided solid [tetrahedron]. The sphere which is described about this will be that of Jupiter. Let Jupiter's orbit be circumscribed by a cube. The sphere described about this will be that of Saturn. [...] Now, place a twenty-sided figure [icosahedron] in the orbit of the Earth. The sphere inscribed in this will be that of Venus. In Venus' orbit place an octohedron. The sphere inscribed in this will be that of Mercury.[105]

Kepler obsessed over this for the rest of his life. About this experience he wrote, "It will never be possible for me to describe with words the enjoyment which I have drawn from my discovery. Now I no longer bemoaned the lost time. I no longer became weary at work; I shunned no calculation no matter how difficult."[106] However, he could not resolve the platonic solids with the exact known orbital values. Kepler was the most prominent astronomical theoretician of his day. He is famous because he developed three mathematical laws that accurately describe gravity and planetary motion. Is it possible that his notion about the Platonic solids was merely a veil for something else? He was never able to resolve the real observations of the planetary positions with the Platonic solids. It is said that as a true scientist, he had to give up his treasured beliefs to resolve with the facts, even though the nested platonic solids do actually approximate the orbital spheres (within 10%). Consequently, he became the very first documented human to truly understand the orbital motion of the planets. In geometry, Kepler glimpsed perfection. He wrote, "Geometry existed before the creation. It is coeternal with the mind of God.

---

[105] Max Casper, *Kepler*. (New York: Dover Publications, 1959, 1993) p.63.
[106] Ibid. p.63.

Geometry provided God with a model of the creation. Geometry is God himself."[107]

Kepler lived in a time when the teaching of a sun-centered solar system was not only against the law, but was also punished with torture and death. The persecution of a heliocentric cosmology, came from both the medieval Roman Catholic Church and the scientifically ignorant Protestant Christian leaders as well. What is significant to the thesis of this book, are the specific geometric figures Kepler associated with Earth and Venus. We might posit, *that Kepler had essentially the same knowledge as that of the astronomer Flegetanis in Wolfram von Eschenbach's Parzival.* As had Flegetanis, did Kepler also see, "hidden secrets in the constellations?" Just like Flegetanis he certainly was able to, "define for us the recession of each planet and its return, and how long each revolves in its orbit before it stands at its mark again." Is it possible Johannes Kepler knew about the tracings of the orbital conjunctions of Earth and Venus shown in this book? Did he see, "the Grail whose name he [Flegetanis] had read clearly in the constellations," and that he hid what he knew behind the symbolism of the Platonic solids he imagined in the planetary orbits?[108]

In his writing Johannes Kepler compared the famous theorem of Pythagoras to a measure of gold and the Divine Ratio to a "precious jewel". This may indicate that he understood as well the Phi ratio's connection to the Grail symbolized as a gem, stone, or rock in Parzival. Kepler may have revealed the secret of the Grail by *associating the octahedron inside the orbit of Venus and the dodecahedron outside the orbit of Earth and the icosahedron between Earth and Venus.* Kepler's German intellectual background and

---

[107] As related by Carl Sagan in the *Cosmos* documentary, "Harmony of the Worlds."

[108] Eschenbach, *Parzival.* p.232.

language could have made him familiar with Wolfram von Eschenbach's Parzival and Wartburg War. It is possible Kepler recognized the Phi ratio as the jewel struck from Lucifer's crown that fell to earth and was refashioned into the cup of Christ. Kepler was a devout Lutheran and in 1521 the Protestant heretic Martin Luther had been given asylum at the same Wartburg Castle in Germany where Wolfram had engaged in poetic battle with Heinrich von Waldeck. The resulting poem was called the Wartburg War, which speaks of the jewel Saint Michael removed from Lucifer's crown. Wartburg Castle was possibly associated with esoteric societies preserving esoteric knowledge over long centuries. Wartburg is also associated with the "Legend of Tannhauser" which preserves the myth of the Goddess Holda, also known as the "Venus of the Teutons," who was literally (as in *literature*) mythically forced underground and into a cave in this legend.

The famous Oxford theoretical physicist Stephen Hawking, in his book "On the Shoulders of Giants" mentions, "Kepler was so obsessed with measurements that he even calculated his own gestational period to the minute – 224 days, 9 hours, 53 minutes. (He had been born prematurely.)"[109] Note that this is not the normal duration of a human gestation. It is surprising that Stephen Hawking did not notice or comment on the fact that *Kepler's proclaimed personal gestational period is exactly the time of the orbit of Venus* as measured in Earth days. This may be a clue to the importance of the orbit of Venus, *because Kepler pronounces his "birth" or perhaps his inspiration or even his original epiphany comes from the orbit of Venus!*

---

[109] Steven Hawking, *On the Shoulders of Giants.* p.627

By the circumstances surrounding a major event in Kepler's life and in history, more circumstantial evidence accrues. This was the appearance of a supernova in 1604. It was observed all over the globe, but Johannes Kepler observed it and wrote a book about it. It subsequently became known as "Kepler's Supernova" or "Kepler's Star." In the fall of 1604, there was a conjunction in the western sky of Mars, Jupiter, and Saturn. The supernova blazed forth within this tight grouping of planets, and was said to have been brighter even than Jupiter. The supernova was visible to the naked eye in the constellation of Ophiuchus for two years. There was great excitement over this event, especially in the wonder that the new star appeared at the same time as the planetary conjunctions, and actually appeared within the very conjunction of those planets. It was widely regarded as a great portent of some momentous change in the world and some predicted the beginning of a new age of enlightenment, which is in fact, about when the historic "Enlightenment" or "Age of Reason" began. While the event itself may not have brought about a change in consciousness or a new age, the attention given to it and the belief in it as a great portent, could have been a catalyst for change. Certain aspects surrounding the year 1604, particularly the retroactive assigning of mythological significance to the year of Kepler's Supernova, may be evidence connecting Kepler to a famous secret society of his day known as the Rosicrucians, a "College of Invisibles," that generated the myth. The year 1604 is a year special to the esoteric Rosicrucians who claim 1604 is the year that the tomb of Christian Rosenkreutz was opened, and that the ancient wisdom traditions were being re-released into Europe after this time. This exploding star of 1604 was first observed, by some accounts, by an imperial official on October 10th and 11th. Other accounts place the first sighting on October 9th. There is a holy significance to the date, which could not have been missed by the esoteric community of that time, because the date references a chapter and verse in the Book of Revelation. The Rosicrucians caught the imagination of Europe with the anonymous publication of three manifestos in the form of pamphlets. These were the Fama Fraternitatis in 1614, the Confessio Fraternitatis in 1615, and The Chemical Wedding of Christian Rosenkreutz in 1616. The esoteric societies would have viewed the coming year 1618 as very significant. That year was the eagerly awaited turning point of the new enlightened age and a very special year that would never occur again with the current calendar. Kepler's final and greatest work, "The Harmonies of the World", was finished in May of 1618. This may have been one of the firebrands starting the very long war that began that very year. The year 1618 was important because, on the one hand, the digits 1618 include the very numbers that express the Phi ratio (1.618), on the other hand esoteric societies may have believed in the reestablishment of the new reformed Christian church upon the original rock of science, as is implied in the numerically significant Matthew 16:18. Although changes were in the wind, there was an initial nasty backlash against the Enlightenment called the Counter Reformation. The result of this was the Thirty Year War of religion between Protestants and Catholics from 1618 to 1648.

In his "Mysterium Cosmographicum," Kepler described the Earth orbit enclosed by a dodecahedron, with an icosahedron inside it enclosing Venus's orbit and an *octahedron inside Venus*, enclosing Mercury's orbit. As the icosahedron is to the dodecahedron so is the orbit of Venus to the orbit of Earth. Note that only Venus generates the *eight-pointed star* so placing the octahedron inside the orbit of Venus suggests Kepler is indicating this fact of the 365 day orbital conjunctions of Earth and Venus. However, both Venus and Earth trace the pentagram and both the dodecahedron and the icosahedron contain the Phi ratio. Therefore, Earth enclosed by the dodecahedron contains the Phi ratio and Venus enclosed by the icosahedron also contains the Phi ratio. Imagine how similar these look compared to the icosahedron inside the dodecahedron. The vertices of the icosahedron are centered in the pentagonal faces of the dodecahedron. If you visually trace those vertices in your mind's eye, the pentagram emerges from the structure. The proportions of the icosahedron compared to the dodecahedron are very close to the proportions of the Venus pentagram compared with the Earth pentagram. It is possible this is what inspired Kepler's epiphany on July 19th 1595, and inspired his book "Mysterium Cosmographicum."

The Great Work of esotericism is the alchemy of the ages. On the small level, it is about the refinement of individuals, to turn the leaden humanity into golden humanity. On the larger scale, it is about improving and evolving the ever-unfinished pyramid of civilization. See an image of this unfinished pyramid on the reverse of the Great Seal of the United States of America and the dollar bill. True teachings are often hidden behind arcane symbolism. The Great Work that has consumed illuminated souls throughout all eternity has been the greatest of magical conjurations, which is the creation of civilization itself. To bring about a great change, such as the evolution of civilization from the Medieval age into a new enlightened era, takes a great deal of effort on the part of many people to educate the world into new ways of thinking. During the Elizabethan era this Great Work was known in esoteric circles as "The Great Instauration of

Light." In specific ways, occultists have been able to transcend time and communicate with the future using certain markers left for seeking individuals to find like treasures from the past. Knowing what has been explained up to now in this book and this theory of the Grail, *one may look for evidence of the activity of esoteric societies in the year 1618.*

CHAPTER 21

Universal Reformation of the World

For I the Lord your God am a jealous God, visiting iniquity of the fathers upon the children to the third and fourth generation of those who hate me, but showing steadfast love to thousands of those who love me and keep my commandments.
Exodus 20:5-6 RSV

Through all eternity illumined souls have been engaged in the "Great Work". They have left behind evidence to show us their presence. One example is the illustration called "The College of the Fraternity," which was published in 1617 in the "Speculum Sophicum Rhodo-Stauroticum" by Theophilus Schweighardt. The title means, "The Mirror of the Wisdom of the Rosicrucians."[110] In 1589 Kepler attended the University of Tubingen. In 1597 Kepler published "Mysterium Cosmographicum," about a mystery of the

[110] For a discussion of the symbolism see "The Rosicrucian Digest" Volume 82, Number 1, 2004.

Cosmos. Key features of the book include heliocentricity, geometric platonic solids (which hark back to the Neo-Platonic Alexandrine schools), the "harmony of the spheres," and in which Kepler refers to the Divine Ratio as a "precious jewel." In the year 1600 the scholar Giordano Bruno is gagged to prevent him from speaking and burned at the stake in the Campo de' Fiori, a market square in Rome. His crimes included the propagation of specific ideas including; heliocentricity, hermeticism, an infinite limitless universe, the idea that all stars are suns like our own, and that there will be life with intelligence elsewhere in the universe.

This act of violence may have been designed to signal a threat across Europe toward all intellectuals and freethinkers, possibly targeting Kepler specifically. In 1604 a totally unexpected event occured. A supernova blazes forth in the sky and is hailed as a harbinger of a New Age of Man. Kepler became a popular celebrity astronomer and astrologer and wrote extensively on the meaning of the new star. Common people in the streets discussed "Kepler's Star." The year 1604 is also the year King James 1 commissioned his famous English translation of the Bible, which was in its first printed edition seven years later.[111] In 1614, 1615, and 1616, anonymous Rosicrucian pamphlets are circulated in Europe and create a furor for the Reformation of Society to bring about a new Age of Reason. In 1614 the first manifesto was entitled, "Universal and General Reformation of the Whole Wide World." About these manifestos, according to the Rosicrucian Digest (Volume 91 Number 2 2013), "Most present-day experts agree that they were not the work of one person, but rather of a small group of students and scholars living in Tubingen, a university town in

---

[111] Butler, *Hiram Key Revisisted.* p.191.

Wurttemberg, Germany. This group was called the 'Tubingen Circle'." It is possible Johannes Kepler knew members of the Tubingen Circle, as well as other Rosicrucians and Freemasons of his Age including his contemporary Francis Bacon, one of whose books is, "The Great Instauration [of Light]". When those who were in the know, publicly spoke about "Kepler's Star" they may not have been necessarily talking about the supernova. Perhaps it was also a reference to another geometric star Kepler discovered in the heavens above? If true and if he had been more direct about his mystery of the cosmos he may have been arrested, tortured, and condemned like Giordano Bruno. *Perhaps the real secret about the year 1604 is that it was the year in which the secret of the of the Holy Grail was being propagated by the "invisible college" to those with the eyes to see and ears to hear?*

In the social and political climate of his day, this would be religiously explosive. Today, one may hope it will be merely controversial. You cannot forever eradicate the Truth, especially when it is an astronomically large signpost in the sky, *as large as the planetary orbits are wide!* Knowing the importance of the Phi ratio, the year 1618 (1.618) would have been highly significant as the fulcrum or turning point for the beginning of the New Enlightened Age of Man; in which there is no longer any forbidden truth. Evidence for this, is in the illustration of the College of the Fraternity showing the year 1618 prominently in its banner. The invisible college is in a mobile fortress with letters flying as though birds. Note that it is recorded that an imperial official first recorded sighting the 1604 supernova on October 10th and 11th, and by some other accounts on the 9th. There is a holy significance to this date that would not have been missed by the esoteric community of that time. Given the portent and claims at that time that a New Age was dawning, could it be any coincidence that October 10th and 11th, indicate a special passage in the Bible? October is the 10th month of the year, but the root "octo" means "eight." *Someone knowledgeable might connect the 1604 Supernova to Revelation 8:10-11.*

> And the third angel sounded, and there fell a great star from heaven, burning as it were a lamp, and it fell upon the third part of the rivers, and upon the fountains of waters; And the name of the star is called Wormwood: and the third part of the waters became wormwood; and many men died of the waters, because they were made bitter. Rev 8:10-11

The closeness of the date to this scripture is such an amazing coincidence, because of the reference to the burning star. It must have been tempting to think this supernova was a portent sent by God on the specific date to corresponding to this quote. Was it God's will that the Blazing Star or pentagram should be restored to its place in the heavens above? In reality, the esoteric initiates of the day could not have missed the significance of the date and would have opportunistically exploited the event. If the date was really October 9th, they may have retroactively amended the date to the 10th and 11th to draw attention to this scripture and to you today. After all, astronomical first sightings are always reported on the first single day and not on two days. In addition, this would have been more apparent to people in Kepler's time, because of the celebration of the feast of Michaelmas. Michaelmas is a celebration of the conquering of Lucifer and currently observed on September 29th. But in olden times Michaelmas was celebrated on October 10th and 11th, and this is called "Old Michaelmas." To the initiates of that age those dates would have been an obvious connection to scripture. (Rev 8:10-11) The very association of this scripture with Kepler's Supernova as the restoration of the pentagram, may well have been the reason that the Roman Church moved the celebration of Michaelmas to September 29th, to make the connection to 1604 and the Rosicrucians less obvious. Recall the account of the astronomer Flegetanis seeing the name of the Grail in the constellations, and that the Grail is the "stone that fell from heaven" in Wolfram von Eschenbach's "Parzival". Peter the Rock, is represented by the phi ratio, which exists infinitely in the pentagram. This same rock was the biblical "stone rejected by the builders," which is also the "star cast out of heaven." This rock is the jewel, or Grail Stone, struck from Lucifer's crown in the German medieval poem "Wartburg War," by Wolfram von Eschenbach. In the poem, the crown is not described but it is likely the same thing as what, "the heathen Flegetanis saw – and he spoke of it reverentially – hidden secrets in the constellations, he declared there was a thing called the Grail whose name he had read clearly in the constellations."

Lucifer/Venus is the name of the Grail in the constellations. The crown itself is the pentagram. The jewel or rock that fell to earth and was re-fashioned into the cup of Christ and found by Parzival is the Phi ratio from the 36° tip or point of the pentagram. This is the same rock upon which Christ built his Church (and the gates of Hell shall not prevail against it), in Matthew 16:18 where *Phi = 1.618 = Matt 16:18 = Holy Grail = rock = Cup of Christ*. This is a jewel or stone of astronomical knowledge about life and Venus that fell to earth. In time it was then refashioned by initiates of the esoteric mysteries into the Grail or cup of Christ. Others have written that the Grail is probably of pagan origin, and here it is plain that the goddess Venus, and the Phi ratio, and this astronomical knowledge were refashioned into the cup of Christ. It is this knowledge about Venus that is meant in Revelation in the words attributed to Jesus.

> And I will give him the morning star. He who has an ear, let him hear what the [Holy] spirit says to the churches. Rev 2:28-29

The result of having this knowledge could turn you into a "neutral angel," or at least perhaps one might lose prejudice against the pentagram star. When right and left or good and evil become relative to your point of view, you can see both sides of an argument and therefore the Grail knowledge causes neutrality, which may bring the enmity of God, or at least enmity from self-righteous religious dogmatists.[112] Recall that in Parzival, Wolfram initially suggests that the neutral angels and the Grail family were forgiven by God, and then at the end of his tale he retracts this, for God is constant in persecuting a balanced viewpoint. But if the stone rejected by the builder now becomes the headstone of the corner, then we may surely say, "It is the Lord's doing and marvelous in our eyes."[113] For a great evil may be undone. The Blazing Star cast out of heaven may be restored to its rightful place in heaven. Those who were embittered by this desecration of what was once holy were the scientists and those who upheld Truth. As the scripture says, "and many men died [...] because they were made bitter." The German philosopher, Friedrich Nietzsche identified them as, "the *good old* philologists and doctors of the Alexandrine schools."[114] It was on them that the war against astronomy, science, and reason was fought. Later in history, they were called witches, warlocks, and sorcerers. (In fact the conical hats of fictional wizards and witches indicate the conic sections of the advanced ancient geometricians.)

---

[112]Rev 3:15-16 "I know your works; you are neither cold nor hot [...] So, because you are lukewarm, and neither cold nor hot, I will spew you out of my mouth."

[113]Ps 118:22-23

[114] Friedrich Nietzsche, *The Antichrist: A Criticism of Christianity*. (New York, NY: Barnes & Noble, Inc., 1895, 2006). p.52.

Restoring the Blazing Star to its rightful position in the stars will heal the rift between science and religion. By embracing Truth, we heal ignorance, which is called the most loathsome of diseases. The astronomical knowledge of the tracings of the planetary conjunctions of Earth and Venus, and its mystical connection with life on Earth joins science and religion in balanced harmony. It reconciles superstition into a *right relationship with reality.*

CHAPTER 22

The Stumbling Stone

Jesus said, "Show me the stone
which the builders have rejected.
That one is the cornerstone."
The Gospel of Thomas (66)

The Holy Grail is thought to be the cup Jesus drank from at the Last Supper, but this is just one of a number of metaphors for what is the greatest mystery of the last 800 years. The most significant of the Grail romances is "Parzival," written between 1195 and 1216. By examining the unique symbolism in this work by Wolfram von Eschenbach, we may come to a solution of the mystery of the Grail. The secret veiled behind the symbolism is related to an astronomical phenomenon involving the Divine Ratio and mystically connected to the genesis of life on Earth. The Knights Templar are common to all the Grail stories as the defenders of the Grail. In 1307, the Pope and the King of France persecuted the Templars. Books such as "Born In Blood – The Lost Secrets of Freemasonry" by John J. Robinson suggest that the Templars went underground after being persecuted on

Friday, October 13, 1307. This is the origin of the superstitions about Friday the 13th. Some Templars then joined with Robert the Bruce in Scotland and fought at the Battle of Bannockburn. They resurfaced centuries later as the Freemasons who were instrumental in founding the United States of America based upon high ideals protecting human rights. Within "Morals and Dogma" by Albert Pike, the highest published mystery of initiation is the 32nd degree called, "Master of the Royal Secret." The graphic for this degree shows a being labeled "Rebis" with two heads, male and female, holding a compass in the male hand and a square in the female hand. There are symbols of the sun and moon and planets. The Rebis stands on Typhon, the Dragon of Ignorance, which is laying upon a tracing board. *The Rebis is known as the end product of the alchemical Magnum Opus or Great Work.* The thesis leading to this book, was that if the Freemasons inherited the knowledge of the Knights Templar, they would likely preserve the knowledge of the secret of the Grail. Beginning on page 841, of "Morals and Dogma" you will find this key text:

> ***Measure a corner of the Creation***, and multiply that space in proportional progression, and the entire Infinite will multiply its circles filled with universes, which will pass in proportional segments between the ideal and elongating branches of your Compass. [...] All hypotheses scientifically probable are the last gleams of the twilight of knowledge, or its last shadows. ***Faith begins where Reason sinks exhausted***. Beyond the human Reason is the Divine Reason, to our feebleness the great Absurdity, the Infinite Absurd, which confounds us and which we believe. ***For the Master, the Compass of Faith is above the Square of Reason; but both rest upon the Holy Scriptures and combine to form the Blazing Star of Truth.*** Magic is the science of the Ancient Magi: and the Christian religion, [...] reveres those Magi who came from the East, guided by a Star, to adore the Saviour of the world in His cradle.

> Tradition also gives these Magi the title of "Kings;" because initiation into Magism constitutes a genuine royalty; [...] The Star which guided them is that same Blazing Star, the image whereof we find in all initiations. To the Alchemists it is the sign of the Quintessence; to the Magists, the Grand Arcanum; to the Kabalists, the Sacred Pentagram. The study of this Pentagram could not but lead the Magi to the knowledge of the New Name which was about to raise itself above all names, and cause all creatures capable of adoration to bend the knee. [...] ***When Truth comes into the world, the Star of Knowledge advises the Magi of it,*** and they hasten to adore the Infant who creates the Future.[115] (italics and bold added)

Faith begins where reason sinks exhausted, because attempts to understand why the Phi ratio exists in both the planetary orbits and also life on Earth and have failed, acquiescing to human feebleness before Divine Reasoning. *Holy Scriptures combined with the science of astronomy, forms the Blazing Star of Truth and this is the most hidden secret of the Holy Grail.* In "Parzival" the Grail is uniquely a stone. Referencing a stone or rock, is the key to finding illuminating biblical passages, especially of a "rejected stone." Wolfram claims he based his book upon privileged information he received from a scholar from Provence, France named Kyot. Kyot himself received his knowledge from a manuscript found by him that was laying neglected in the corner of a library in Muslim possessed Toledo, Spain. The author of the text was Flegetanis, an astronomer who wrote of the marvels of the Grail. The following passage sets Parzival apart from all the other versions of the Grail story because it is explicit in providing concrete clues leading to the Grail.

> For the infidel Flegetanis was able to define for us the recession of each planet and its return, and how long each

---

[115] Albert Pike. *Morals and Dogma of the Ancient and Accepted Scottish Rite of Freemasonry*. (Montana USA: Kessinger Publishing Company, 1871). p.841

> revolves in its orbit before it stands at its mark again. All human kind are affected by the revolutions of the planets. With his own eyes the heathen Flegetanis saw - and he spoke of it reverentially - hidden secrets in the constellations. He declared there was a thing called the Grail, whose name he had read clearly in the constellations. A host of Angels left it on Earth.[116]

Wolfram's passage reveals the secret that will make it possible to find the Holy Grail. The first step would be obtaining sufficient astronomical knowledge of the planets to have the ability to mark or trace planetary orbits around the sun. The theme developed by Wolfram von Eschenbach, is that during the mythic war of the Angelic Rebellion a host of angels brought the Grail to Earth from heaven and protected it until it became the responsibility of worthy humans. According to Wolfram:

> When Lucifer and the Trinity began to war with each other, those who did not take sides, worthy, noble angels, had to descend to Earth with that stone which is forever incorruptible.[117]

Neutral angels brought the Grail, which is called an incorruptible stone, to Earth to keep it safe during the time of war. On every Good Friday, a dove, depicting the Holy Spirit, flies down from heaven bringing a small white wafer to the stone and then returns to heaven. The wafer gives the stone its magical powers. The Grail functions similarly to the ancient Greek mythical concept of a cornucopia or horn of plenty, producing food and also birds, land animals, and fish. Wolfram narrates:

> Every Good Friday, as I say, the Dove brings it to the stone, from which the Stone receives all that is good on Earth of food and drink, of Paradisal excellence – I mean whatever

---

[116] Wolfram Von Eschenbach, A. T. Hatto (Translator). *Parzival*. (London, England: Penguin Books Ltd, 1980). p.232

[117] Ibid, p.240

> the Earth yields. The Stone, furthermore, has to give them the flesh of all the wild things that live below the aether, whether they fly, run or swim - such prebend does the Grail, thanks to its indwelling powers, bestow on the chivalric brotherhood.[118]

In 1207, Wolfram von Eschenbach was in the court of Hurmann, Landgrave of Thuringia at Wartburg Castle near Eisenach in central Germany. There Wolfram engaged in a poetic battle with another famous poet Heinrich von Waldeck. The result is the epic poem "Wartburg War." In strophe 143, we find mention of the Grail with the tale of the Angelic Rebellion.

Shall I then bring the crown,
that was made by 60,000 angels?
Who wished to force God out of the Kingdom of Heaven.
See! Lucifer, there he is!
If there are still master-priests,
Then you know well that I am singing the truth.
Saint Michael saw God's anger, plagued by this insolence.
He took Lucifer's crown from his head,
In such a way that a stone jumped out of it,
Which on earth became Parsifal's stone.
The stone which sprang out of it,he found it,
he who has struggled for honor at such high cost.[119]

Once a year on Good Friday, a dove, depicting the Holy Spirit, flies down from heaven bringing a small white wafer to the stone and then returns to heaven. The wafer gives the stone its magical powers to produce food and drink and whatever the Earth yields. Lucifer has associations that many people are unfamiliar with. For instance, the name Lucifer means light bringer (luce light + ferre bring). There is a cave in the south of France in

---

[118] Ibid, p.240

[119] Karl Simrock. *Der Wartburgkrieg.* (Cotta, Stuttgart, Augsburg 1858. OCLC 5363166), a German translation of the medieval Codex Manesse, Janaer Leiderhandschrift and translated to English.

Saint-Baume that in Roman times was the center for worship of the goddess Diana Lucifera the light bringer or Illuminatrix. Lucifer is also the Latin name of the second planet from the sun; more commonly called Venus after the goddess of love.

The English language evolved in the historical period following the Norman Conquest of Britain in 1066 A.D. As the Normans assimilated themselves into Britain's population, the French language of the Normans and the Germanic tongue of the Britons merged to form the English language. The English alphabet was designed by the Normans in the generation following the Norman Conquest. They adopted the Roman alphabet and added three new letters to it - JUW.[120] Because of the addition of the new letters JUW in specific placement, the English alphabet contains two very significant mathematical constants encoded within it, being the values of Pi and Phi. It should be obvious that having two mathematical constants encoded to four decimal places cannot possibly be by chance. You may reveal the encoded Pi and Phi ratios by arranging the characters in a circle and separating the symmetrical characters from the unsymmetrical ones. Some group of cryptographers went to great lengths to encode these values. What was encoded in the alphabet, the circle (Pi) and pentagram (Phi) combine to form a pentacle. [See Appendix for Details] *That a pentacle is encoded in the English alphabet is evidence of illumined minds and esoteric society acting out an inter-generational plan and opening a special channel to communicate across the centuries for those with the "eyes to see and ears to hear," who know where to look and what to listen for!*

The event called the Pentecost fifty days after resurrection similarly encoded a pentacle (50='5'+'0'), a

[120] William Eisen, *The Universal Language of Cabalah.* (Marina Del Rey, CA: DeVorss & Company 1989) p.21.

symbol that was identified with evil in the Dark Ages, and which was associated with black magic, the Antichrist, and 666. The event of the Pentecost describes a mighty wind from heaven that fills the house where disciples are sitting and tongues of flame come to rest on their heads. The tongues of flame on the head is the same in meaning today of a light bulb going on over our heads. It is a metaphor for an epiphany or revelation of knowledge. At the event called Thc Epiphany all can speak to one another in many languages (tongues) and understand. This reverses the idea of the Old Testament Tower of Babel, where God inflicts mankind with a confusion of languages; which would otherwise allow the human race to make great achievements. (Gen 11:1-9) Speaking a common language implies mankind working toward common goals. It may be that what the Pentecost symbolizes and what transpired two millennia ago was some knowledge which transcended language, tribe, religion, and national borders, and was something evident and understandable by all, like a supernova appearing in the sky. (Acts 2:1-14)

Manly P. Hall considered King Arthur to be the Grand Master of a Christian-Masonic brotherhood and that the historical Arthur became confused with the allegories and myths of his esoteric mystical order.[121] Imagine that the historic Jesus was similarly confused with the symbolism of the stone as his teaching in the following passage.

> 1 Therefore, laying aside all malice, all deceit, hypocrisy, envy, and all evil speaking, [...] 4 Coming to Him as to a living stone, rejected indeed by men, but chosen by God and precious, 5 you also, as living stones, are being built up a spiritual house, a holy priesthood [...] 6 Therefore it is also contained in the Scripture, "Behold, I lay in Zion A chief cornerstone, elect, precious, And he who believes on Him will by no means be put to shame." 7 Therefore, to you

[121] See chapter "Repanse do Schoye".

> who believe, He is precious; but to those who are disobedient, "The stone which the builders rejected Has become the chief cornerstone," 8 and "A stone of stumbling And a rock of offense." *They stumble, being disobedient to the word, to which they also were appointed.* 1 Peter 2:1-8 NKJV

In the sense of rejected knowledge, "being disobedient to the word" is not an abstract but rather a literal meaning, meaning rejection of knowledge; rejection of something truthful, but which some may not understand, which is lost in translation, or which some find offensive. This idea subtly changes the image of the passage above like in a mirror. The association of Christ with the pentagram star, the planet Venus, and Lucifer has clearly been rejected. This very rejection makes it a candidate for "the stone rejected by the builders," or the rock of scandal upon which men trip and fall.[122] Consider the above quote in conjunction with the following parable.

> Give not that which is holy unto the dogs, neither cast your pearls before swine, lest they trample them under their feet, and turn again and rend you. Matt 7:6

Who are those who stumble upon the "word" to which they have been appointed? They are supposed to be those who have been appointed to the task and who are as "living stones" and to testify on behalf of the stone (rock) that was "rejected indeed by men, but chosen by God and precious" like a gem; and they are those who gain power and prestige being "a holy priesthood." In John 1:1, "In the beginning was the Word." This is where the disconnect is, where the knowledge and identity of the person are confused. The "word" is the Logos, a common Greek concept; which is the spirit of Reason and Truth; it is acceptance of "aletheia" naked Truth, that which is not

---

122 1 Peter 2:8, Psalms 118:22-23.

concealed and out in the open, and a mind which is not closed. The usual interpretation of the passage above, is the dogs and swine are base people, but there is a more sophisticated interpretation. There is more to expound upon, perhaps in another book; but those who stumble on this stone, those who reject the stone and therefore disobey the Word (Logos – logic and reason), and who will turn and bite you – *are religious dogmatists*!

Consider this verse from the Gospel of Thomas, "Jesus said, 'Show me the stone which the builders have rejected. That one is the cornerstone.'" (66)[123] The medieval Church made this forbidden knowledge, but doing so is antithetical to the teaching of Christ, "*For there is nothing hidden which will not be revealed, nor anything that has been kept secret that should not be revealed.*" (Mk 4:22) Perhaps if someday this becomes generally known and we restore the pentagram to its rightful place in heaven, then the following passage will reverberate. "The stone which the builders rejected is become the head stone of the corner. This is the LORD's doing; and it is marvelous in our eyes. (Ps 118:22-23)

Here is revealed the essential secret of the Grail that was "rejected by the builders." The interpretation of the Grail stone presented here will surely cause some to trip and stumble. The stumbling stone is an inconvenient Truth, which violates some rule of mythical coherence. Yet this truth may literally save the human race if it may lead us to a New Earth and a new beginning!

---

[123] Barnstone. *The Other Bible.* p.305.

CHAPTER 23

The Signpost of God

This is articulus stantis et cadentis ecclesia
—that article, with the admission or the denial of which
the church either rises or falls;
"the main hinge on which the door of salvation turns."
Matthew Henry's Commentary on Matt 16:18

It is possible that the theory of the Grail explained here provides the only concrete reason as to why the medieval Catholic Church would have wanted to prevent the accurate development of astronomy and outlawed heliocentricity as a heresy punishable with death. The usual reason given to explain this phenomenon of propaganda and enforced ignorance, was that it did not put the earth at the center of God's creation. But the real reason may have to do with the orbital harmonics of Earth and Venus. The reader will find information that associates the pentagram with an astronomical phenomenon in "Dictionary of Symbols," (1994) by Carl G. Liungman.

> From the sixteenth century until the eighteenth century the pentagon, as a structure closely associated with the

> pagan goddess through the celestial movements of the planet Venus, was common as a basic design for fortresses.[124]

Here is an incredible reason that, for instance, the command center of the United States military is a Pentagon complex in Arlington, Virginia. Carl also wrote about the pentagram that, "The sign is the structure that results if one plots the movements of the planet Venus *as seen from the Earth in the zodiac*."[125] This is not true as any attempts to reproduce the pentagram this way will fail such as when you observe Venus over time in the zodiac using astronomical software. You must instead employ the heliocentric perspective to produce the pentagram. Consider the literal perspective of looking at the heavens from the surface of the Earth or the alternative viewpoint of the sun. The different perspective is very large seen from these different locations. [See Figure "The Error of Geocentric Perspective"] Therefore, if the astronomer Flegetanis gave instructions on seeing hidden secrets in the constellations" then the geocentric perspective will never show it. An entirely different view of the heavens will be the result.

The meaning of the crown of Lucifer/Venus referenced in "Wartburg War" and the "Secret Teachings of All Ages" is obvious if you know that since about 4,000-3,200 BC the planet Venus has been associated with the pentagram star. The archaeological evidence linking Venus to this icon in the distant past are proof that the exact astronomical knowledge of the forbidden Copernican cosmology was present at the dawn of civilization. It is this knowledge that may be the essential secret and truth preserved in the symbolism of the Holy Grail of Christ. It is even arguable

---

[124] Carl G. Liungman. Dictionary of Symbols. (New York, London: W.W. Norton & Company, 1991) p.314.

[125] Ibid, p.43.

that what some people consider the time of creation; is really the age when human civilization understood what the mythologist Joseph Campbell in "The Flight of the Wild Gander" described as, "the recognition, some time in the fourth millennium B.C. of the orderly round-dance of the five visible planets and the sun and moon through the constellations of the zodiac."

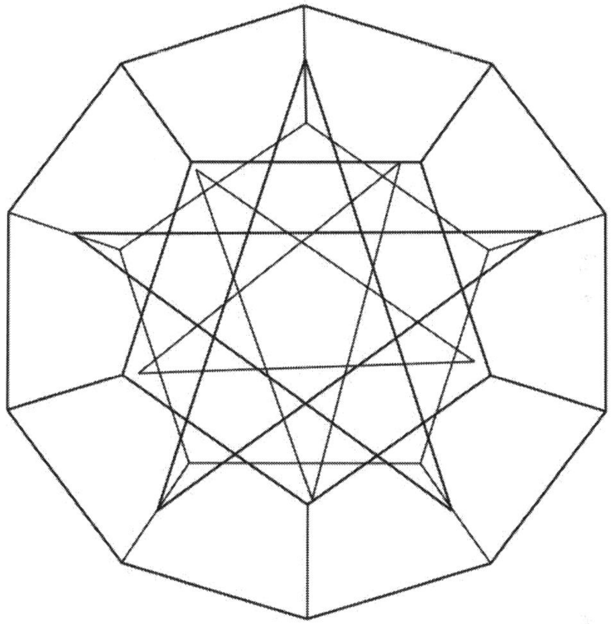

Figure 12 The Signpost of God

Venus traces a pentagram in the zodiac every eight years. If Venus traces a pentagram in the heavens above, then the Phi ratio exists in the planetary orbits above, as well as in life forms here on Earth below - *yet at this time of writing these tracings are not in astronomical text books as common knowledge. This is absolutely astounding! How did this planetary dynamic evolve?* No other Grail solution so perfectly explains Leonardo DaVinci's enigmatic John gesture found in his last painting called "John the Baptist" who points upward to heaven with his index finger and has an enigmatic smile. Other authors have noted this painting may be associated with the Grail secret, but without knowing what this significance might be. The portrait of John the Baptist seems to be saying, "*If you want to see evidence of the divine then look to the heavens - there is an astronomically huge sign there that is like a beacon indicating life is here in this solar system.*"[126]

However you might interpret this, Leonardo DaVinci was making a very special gesture – possibly to the Church of his day. Whoever might be truly serious about studying this would need to acquire a working knowledge of the astronomical sciences. Anyone with a background in math and software development might study the authoritative work, "Astronomical Algorithms" by Jean Meeus.[127] *The Grail as the rejected pentacle (rejected stone) is literally in the heavens above.*

In the "Timaeus" Plato wrote, "and there is a fifth figure which is made out of twelve pentagons, the dodecahedron – this God also employed in the construction of the universe." In his "Mysterium Cosmographicum" Johannes Kepler wrote, "The Earth is the measure for all other orbits. Circumscribe a twelve-sided regular solid [dodecahedron] about it [...]." The image included in this chapter is a two-dimensional projection, of a three-dimensional computer model, tracing the 584-day orbital conjunctions of Earth and Venus. Projected within this model is a dodecahedron with an exact radius of one astronomical unit, which is the distance of Earth's orbit. It is striking that the harmonics of the Earth orbit will intersect the center point of the dodecahedron face, and

[126] The Appendix of this book contains an astronomical program to calculate the Earth and Venus orbits for anyone so inclined to use for this purpose.

[127] Jean Meeus, *Astronomical Algorithms, Second Edition* (Richmond Virginia: Willmann-Bell, Inc., 1991, 1998).

that the inner orbit of Venus fits the proportion of a single pentagonal face. In homage to H. E. Huntley is this, "A pretty result?"[128] Is it an aspect of Diotima's Beauty given to Socrates and bequeathed to Plato? Is it the pearl that Jesus warned not to cast before swine? (Matt 7:6) It inspires a mixture of archaic emotions. There is surprise that what Plato wrote in "Timaeus," and what Kepler wrote in "The Mystery of the Cosmos" so closely correspond; such that it may be the same exact phenomena as this - that both describe; with about two thousand years intervening. It is surprising that professional astronomers do not appear to be aware of this. For example, the famous astrophysicist Mario Livio, makes no mention of this in his popular book, "The Golden Ratio: The Story of Phi, the World's Most Astonishing Number" (2002); and yet there is curiosity; as to how this phenomenon involving the Golden Ratio can be overlooked. There is wonder that we have chanced upon an unexplored phenomenon; and we may speculate that there may be other significant discoveries to be made here! It appears to represent two Hermetic Principles; one is The Principle of Gender which manifests on all planes of existence; the other is the Principle of Correspondence; as above, so below, as below, so above. In an astronomically large depiction; we see evidence of the pentagonal nature of life in this corner of the Cosmos, and also several magnitudes of existence below on Earth. Kenneth MacLean, in his book "A Geometric Analysis of the Platonic Solids and Other Semi-regular Polyhedra" wrote, "Nucleotides are the basic units of DNA. [...] The central structure, the deoxyribose sugar, is pentagonally

[128] Huntley, *The Divine Proportion*, p.36, see chapter Diotima's Beauty.

shaped. Nature uses geometry in the construction of life!"[129]

Is it humanity's destiny to explore the universe? At this time, scientists are building increasingly sensitive telescopes, and are identifying thousands of planets circling other stars in our stellar neighborhood. Tracing the orbital conjunctions of Earth and Venus, we get geometric shapes that exist in life forms on Earth. What is the connection? Might a similar phenomenon in another solar system lead the way to life elsewhere in the Cosmos?

The secret of the Holy Grail is a teaching about Christ or from a man Jesus that was rejected by the builders of the early Christian Church with the dire consequence of making science and Truth heretical. The Grail is a metaphor of both the Phi ratio (1.618) and the indisputable fact that the pentagram star exists in the heavens above as though an enormous *Signpost of God*, traced by the orbital conjunctions of Earth and Venus every eight years. The connection to Venus is the Grail's connection to goddess worship. The Phi ratio is an essential part of the design of life, and it exists in the human body, which is why it is also called the Divine Ratio.

The pentagram as the symbol of the divine ratio became the most loathsome symbol in Christianity and remains so till this day. It is associated with Satanism and witchcraft *in order to induce fear of it.* The whole "point" of the pentagram, with each tip being 36° degrees; is that it is the geometrical origination point of the Divine Ratio. The mathematical proportion of the length of any two equal length rays originating from a 36° degree point is in Phi ratio to the distance between the two end points. The

---

[129] Kenneth J. M. Maclean, A Geometric Analysis of the Platonic Solids and Other Semi-regular Polyhedra, (2nd Printing – February 2015, Loving Healing Press. 2006, 2007, 2015), p.45

values of 1.618 and 36° degrees and 666 are therefore all representations of the Phi ratio. The angle of 36° degrees is further connected with the "Beast of Revelation" because if you sum up all the numbers from 1 to 36 you get the number of the Beast or 666, where 666 is an encoded form of the pentagram star. This associates the Grail knowledge with the mythical beast, and this reinforces the interpretation of the Grail secret as *possibly the greatest heresy of all time.* This is the most significant interpretation of Revelation 13:18, "Here is wisdom. Let him who has understanding calculate the number of the beast, for it is the number of a man: His number is 666." (NKJV) *The gem that is knocked off the crown and falls to Earth and becomes the Holy Grail is the Divine Ratio and the secret of life itself.*

The underlying reason the Grail is heresy, is because it interprets the rock upon which Christianity was originally built, was not the man Peter; upon which, using the concept of apostolic succession, the Roman Catholic Church claims authority, but rather is the Phi Ratio or 1.618. The rock upon which the church is built comes from Matthew 16:18, "upon this rock I will build my church." Consider the possibility that this verse was placed in scripture precisely where it is because the rock upon which the church was originally built was the Phi ratio. *This is the ratio of life that exists in our physical bodies and multitudinous life forms on the earth, and that ratio 1.618 is contained in the chapter and verse of Matthew 16:18.* It is too significant to be a chance that the verse, upon which the Catholic Church claims its authority, encodes the Phi ratio and references a "rock," or that Parzival symbolizes the Grail as a stone and uses an astronomer (Flegetanis) as the source of Grail knowledge. It exceeds the possibility of the coincidental placement of this biblical verse along with the corroboration of other scripture referencing the morning star or the planet Venus. The Phi ratio fits perfectly the symbolism of the white wafer brought by a dove down from heaven to the stone, which gives it the magical power to produce all life on Earth. It also perfectly fits the passage from Wolfram's "Wartburg War" as the gem being lopped off the crown of Lucifer. To put it differently, this solution to the Grail mystery fits more symbols than any other. More than any other solution to the Grail symbolism, this is the only one, which fits the symbolism of the astronomer Flegetanis clearly reading the name of the Grail in the constellations.

The most common Protestant interpretation of Matt 16:18 is that the rock is Christ himself and not St. Peter. Bible commentator Matthew Henry writes: "*This is articulus stantis et cadentis ecclesia—that article, with the admission or the denial of which the church either rises or falls; 'the main hinge on which the door of salvation turns.'*"[130] The interpretation that the rock is a metaphor for Phi, is sharply in contrast. The Goddess and the Grail symbolized by the pentagram as a burning star, were symbolically cast out of heaven in the book of Revelation. The book of Revelation was written in the late first century and in Rev 8:10-11, a burning star falls from heaven and is extinguished in water, turning the waters bitter. This records the beginning of the persecution of the Goddess and the Pentagram. This passage is almost too coincidentally associated with the date of October 10 & 11 (octo=8) given for Kepler's supernova of 1604, *which does in fact seem like a divine and mystical portent.* Wormwood, also called Artemisia Absinthium and named for Artemis, the Greek goddess of the hunt; is a plant that produces absinthe, a mind altering and alleged brain-damaging alcoholic drink. Within about forty years after the time of

[130] Matthew Henry's Commentary on Matt 16:18.

Jesus, Phi became the "stone rejected by the builders." Along with this, they rejected the Goddess, the pentagram, and the knowledge of the orbital harmonics of Earth and Venus and heliocentricity became a heresy. The symbolism of this passage means this "blazing star" could alter your mind, or damage your perception of religious teachings.

Christianity is monotheistic and portrays God as male. In the pagan and Christian Gnostic perspective, an all-male God concept is out of balance, out of harmony, and a cause of evil in the world. It is only half of the Divine, and it can hardly lay claim to being the better half. This indicates the rejected knowledge of Christ because teachings of Venus or goddess worship conflicted with the patriarchal teachings of Jehovah. Evidence of this can be found in the recently discovered Gnostic gospels, particularly of Thomas and especially Phillip in which we read, "when we became Christians we had both a mother and a father."[131] This implies that the early Christians worshipped both a God and a Goddess or a divine that was androgynous. Knowledge about the Morning Star (Venus) is also the conclusion and the final point of the Bible, being found on the very last page and column.

> 16 I Jesus have sent my angel to testify to you these things in the churches. I am the root and the offspring of David, and *the bright and morning star*. Rev 22:16 KJV

Understanding the Phi ratio as the rock upon which Christ built his church, explains the conclusion of the Bible. *Therefore, by His words the pentagram may be a more fitting symbol for Christ than the cross.* If we are to believe the words recorded in Revelation are indeed the words of Jesus, he declares that he is Venus/Lucifer or they are as

---

[131] Willis Barnstone. (Edited with Introductions). *The Other Bible.* (New York, NY: HarperCollins, 1984). p.88.

one. Such knowledge at the eternal eschatological "end times," could have cosmic significance. Those dogmatic theologians, who in ages past suppressed and persecuted the astronomical knowledge of the orbital harmonics of Venus and the Earth, rejected the stone that was actually meant to be the foundation of the church. The crown of Lucifer being the pentagram star, was cast out of heaven, as was the mythical Lucifer in Isaiah and the Jewel of the Venus/Lucifer crown fell to earth where it was remade into the Grail. The conclusion of the Bible "I am [...] the bright morning star," (Rev 22:16) is a backdoor by which some will exit and some will enter. This is like a theological escape hatch for those who tripped on the Grail stone to become redeemed when the truth inevitably becomes revealed.

The Christian theologians tripped on the stumbling stone of the Grail, because they did not believe they could allow proof to exist of a connection between Christianity and the pagan ancient mystery traditions. If that rejected stone then becomes the head stone of the corner, can you imagine the implications for a New Age? As you digest this idea, and become aware of how difficult it would be for some people if they were told this without suitable preparation, consider how well this fits the symbolism from Parzival, "Every Good Friday, as I say, the Dove brings *it* to the stone". The Dove descends from heaven with a sort of blueprint, from which the stone receives whatever the Earth yields. Then in turn the stone produces all living things on Earth as though a life factory. Then do you think it is suitable, that this "*it*" is found in life on Earth and in your own body, being Phi or 1.618?

Because Christianity was strongly Patriarchal by mutation with Judaism, it denied Goddess worship as a cause of weakening the nation. (Isaiah 14:12) Christian antithesis and claims of exclusivity may deny any commonality with what pre-dates Christianity, and the Grail knowledge proves that Christianity has borrowed from the very distant past and promotes a new Christian paradigm of inclusion of the classical pagan world in its identity. By the words attributed to Jesus, he is the "bright morning star," and perhaps the Bethlehem star seen in the east, is Venus inaugurating a time of enlightenment - or perhaps a New Age?

CHAPTER 24

Sin Most Fiery and Formidable

"What did Jesus say to you?" Thomas said to them,
"If I tell you one of the things which he told me,
you will pick up stones and throw them at me;
a fire will come out of the stones and burn you up."
Gospel of Thomas

One of the sublime interpretations of the meaning of the Grail is the mystical identification with Christ consciousness, which is the very same concept as Buddha consciousness. In this, we find God or Christ in ourselves. One of the expressions of this is, "I am with you always, even unto the end of the world." (Matt 28:20) The knight on the quest for the Grail, after years of seeking outside themselves for the Grail, come to realize after many trials, that they are themselves the "I" or divine vessel receiving direct spiritual contact from heaven, and incidentally bypassing the clergy. This is sharing in the mystical immortality and the understanding of, "before

Abraham was, I am." (John 8:58) According to Joel S. Goldsmith in "The Mystical I":

> It is in this word *I* that you find the entire secret of the spiritual message given to the world by Christ Jesus, a message that is destined to set men free and break from them all shackles and all limitations, that they may live as children of God, completely free, under the domination of no man, under the domination of no circumstances or conditions, and under the grace of God alone. When this is translated into practical life experience, you will begin to understand some of the passages of Scripture that have heretofore been obscure.[132]

The historic problem with this concept is that it implies individual freedom of thought and self-determination as expressed in, "I am the way, the truth, and the life." (John 14:6) One becomes the very incarnation and source of truth and light. The serpent of your soul suffering on the cross of truth whispers, "open your eyes and see." *Then you open other eyes.*

A Gnostic aphorism from the Coptic version of the Gospel of Thomas is, "Split a piece of wood, and I am there. Lift up the stone and you will find me there." (77B) The ultimate Grail secret being the Phi ratio fits this perfectly, because it is found everywhere in life on Earth, from insect life under a rock, to plants and trees, to animal and human life forms. Phi is the divine spark in us all, it is life itself, and humans are the crown of creation and symbolized for excellence by the pentagram star. The Grail is not an artifact or a bloodline of a man; it is the essence of blood itself in the Grail, and a symbol of this knowledge. In other words, the blood in your own veins,

---

[132] Joel S. Goldsmith, *The Mystical I.* (New York, Evanston, San Francisco, London: Harper & Row, 1971). p.39.

with you as the vessel – you are the Grail with the divine spark in you, "the Savior is the Logos made man."[133]

Peter the rock being the Phi ratio of Matthew 16:18 (1.618) rather than Peter the apostle will be too abstract for many to understand. People ignore the blatant mythical similarities between Jesus and Lucifer and the serpent in Eden who wants us to eat from the Tree of the Knowledge so we may open our eyes and be like God. Jesus spoke in code for those with "the eyes to see and the ears to hear." Lucifer wants to be like God, "I will be like the most high." (Isa 14:14) Jesus seems to claim he is God, "I and the father are one." (John 10:30) For this, Jesus is accused of blasphemy. Matthew Henry commented on the heretical nature of Jesus when he wrote, "*in Christ sin is set before us most fiery and formidable.*"[134] This statement in itself is remarkable, and implies an entirely different view on the whole matter; which might take an entire book to explore; suggesting a reverse or even a double reverse psychology in Christianity. The Association of Jesus with the serpent in Eden is made in John 3:14, "As the serpent was raised in the desert, so shall the son of man be raised." It is not a coincidence that the serpent biting its tail (Ouroboros), is a circle and verse 3:14 is the value of Pi (3.14). John 3:14 is a reference to the Old Testament book of Numbers 21:6-9 in which God sends fiery serpents to bite people who will die of the poison unless they look up to a brass serpent that Moses mounted on a wooden cross. This has been used as the justification for using the Old Testament technique of the fire and brimstone "Brazen Serpent." Inject a person with the poisonous venom of Sin, and then require that they look up to the serpent of

---

[133] Pike, *Morals and Dogma – Royal Secret*. p.848.

[134] *Mathew Henry's Commentary*, John 3:1-21, p.1540, on John 3:14 (author's italics).

wisdom crucified on the cross, as an antidote to that poison.

Future research on the Holy Grail may bear fruit in the following areas. One is that the author of the Gospel of Mark, which may be Mark the Evangelist who was the founder of the Coptic Church of Alexandria; which exists to this day; may be a connection to the Greek and Egyptian Mystery schools. Mark is the oldest of gospels and is exactly 666 verses; meaning it is significant to this thesis through that connection to the Mark of the Beast.[135] Another area is the connection of Wartburg Castle to esoteric societies through Wolfram von Eschenbach and to Martin Luther who was rescued to the castle, and there translated the New Testament. Martin Luther used as his symbol a barbed rose similar to that which appears at the center of the historical Round Table of Arthur in Winchester Hall in England.[136] Also the legend of Tannhauser is associated with Wartburg Castle and it is within reason that "Holda" the Venus of the Teutons may have an actual sacred site or cave in the area of Wartburg; as depicted as Venusburg in the Legend and the opera of the same name by Richard Wagner; which includes both Venusburg and Wartburg War in its plot. Also, by examining an original copy of Johannes Kepler's "The Harmonies of the World - Book Five"; published in the year 1618; it may be found that the orbital conjunctions of Earth and Venus are evident; and this may have sparked the Thirty Year War (1618-1638); four hundred years prior to the first publication of this book in the year 2018.

---

[135] See chapter "The Mark of the Beast." Some bible versions omit the following verses 7:16, 9:44, 9:46, 11:26, and 15.28, but the chapter verse counting remains the same. Scholars agree the versions ending with 16:8 (666 verses) are most authentic, given several ending variations of a later date.

[136] It is called Luther's Rose, as described in a Letter from Martin Luther to Lazarus Spengler, July 8, 1530 [WA Br 5:445]; tr. P. T. McCain.

The concepts of the Logos, virgin birth, Immaculate Conception, the Eucharist, and a dying and resurrected savior are among the most common features of the pre-Christian pagan faiths. This is well documented in "The Golden Bough" by James George Frazier, and in "Pagan Christ's" by J. M. Robertson, and in "The World's Sixteen Crucified Saviors or Christianity Before Christ" by Kersey Graves. There are now many books in recent years that connect Jesus with pagan teaching. One is, "Jesus the Magician," by Professor Morton Smith published in 1978.[137] In "The Templar Revelation" by Lynn Picknett and Clive Prince, they discuss the far-reaching implications of a paradigm shift acknowledging his background, from Judeo-Christian to Egypto-Christian. In their conclusion, they remind us of the war against pagans and of the destruction of their temples. They declare, "Yet if we are right, and Jesus himself was a pagan then this Christian fervor was not only once again a denial of common humanity, but also of their founder's own principles."[138] In 1999, Timothy Freke and Peter Gandy published "The Jesus Mysteries," in which they make a lucid argument that the Christian faith was an adaptation from pagan faith and mythology.[139] The Pythagoreans who used the pentagram star as their symbol were pagans in the nicest sense of the word - as erudite scholars, scientists in the modern sense, and philosophers. Here is shown the scriptural basis connecting Jesus with the astronomical knowledge of the Venus pentagram and with the Pythagoreans and the pagan schools.

---

[137] Morton Smith Ph.D., *Jesus the Magician*. (New York, NY: HarperCollins 1978).

[138] Picknett, Prince, *Templar Revelation*. p.364.

[139] Timothy Freke, Peter Gandy, *The Jesus Mysteries, Was the Original Jesus a Pagan God?* (New York: Three Rivers Press, 1999).

It has never before been satisfactorily explained, why it is that in Wolfram von Eschenbach's "Parzival" the Grail is symbolized as a stone. In the highest degree of Freemasonic teaching in "Morals and Dogma" by Albert Pike we find, "For the Master, the Compass of Faith is above the Square of Reason; but both rest upon the Holy Scriptures and combine to form the Blazing Star of Truth."[140] The key to understand the Grail as a stone was to realize that Matthew 16:18, which references a rock upon which Christ's church was built was not a reference to apostle Peter, but to the Phi ratio (1.618). These core ideas went outward like ripples from a stone cast in water and which illuminated so much else. So, what concisely is the secret of the Holy Grail? One may claim only this for sure. This is what Wolfram von Eschenbach believed the Grail to be for which he warned that no one should ask him about directly because he would be forced to lie. The Grail stone is the Phi ratio upon which everything is founded and which is associated with the origin of life on earth. It is possibly the same as the ancient concept of the Gnostic Pleroma, which is symbolic of where God energy or life enters into our Universe. It was the foundational idea present in the origin of civilization in our distant past that represented the attainment of significant astronomical and mystical knowledge. It was like the mustard seed of gospel, which could create a faith to move mountains or find life in the Cosmos.[141] It is the gem cut from the point of Lucifer/Venus's crown (the 36° degree angle of Phi). This is the secret kept for centuries; and feared and forbidden by Christian dogmatists. Venus and Earth trace pentagram stars by their orbital conjunctions and this was enough of a reason to persecute the truth of Heliocentricity for over a Millennia. Nowhere else will

---

[140] Pike, *Morals and Dogma - The Royal Secret.* p.841.

[141] Matt 13:31, 17:20, Mark 4:31, Luke 13:19, 17:6.

you find a more rational explanation for why The Roman Empire purposely retarded the advance of astronomy, which became a contributory cause for the Dark Ages. That some people will find this Truth offensive is the reason the romances say that only the pure in heart will have delight in the Grail, it will not tolerate the presence of the sinful.

As Albert Pike wrote, "God has been pleased to give to Science, which is the reflection of His Glory, the Seal of His Eternity,"[142] which is what may also be called, "The Crown of Creation," or pentagram. Whatever you choose to believe about these matters is up to you. One final thought. This knowledge has in the past been the point where, "Faith begins where Reason sinks exhausted."[143] Yet modern science has advanced greatly in the last 800 years, so the cause and affect explaining the orbital harmonics containing the Phi ratio and also in life on Earth may become a new science. Through the science called spectroscopy we can detect chemical elements billions of light years away. Might this astronomically large *Signpost of God* - lead humanity to *life* elsewhere in the Cosmos?

At the end of the Bible is the curse should anyone alter the message contained there. "For I testify to every man that hears the words of the prophecy of this book, If any man shall add to these things, God shall give to him the plagues that are written in this book: And if any man shall take away from the words of the book of this prophecy, God shall take away his part out of the book of life, and out of the holy city, and from the things which are written in this book." (Rev 22:18-19 KJV) The Grail is not just the rock upon which Christ built his church; it is the foundation stone upon which our civilization was built in about 4,000 – 3,200 BCE. At this point, the last lesson of

[142] Pike, *Morals and Dogma*. p.842

[143] Ibid, p.841

the Buddha comes to mind. When asked what the true way to Enlightenment was, the Buddha replied, “Be your own light, your own refuge. Believe only that which you test for yourself. Do not accept authority merely because it comes from a great man, or is written in a sacred book, for truth is different for each man and woman.” As the instructress Diotima gave Socrates a “beauty” at the end of his initiation; at the end of this current age of humanity, try to imagine this sublime truth radiating to the firmament. May it be, *an enormous Signpost pointing the way to life in the Cosmos?* If Christ had his way he would give the goddess and the Venus star to you, for it is written that he said, “And He that overcomes, and keeps my work till the end [...] I will give him the Morning Star.[144] It is this, which connects Christ to the pagan mystery schools and secret societies, because they and Jesus like Plato’s instructress Diotima, deliver the same beauty at the end of initiation. To life, liberty, and the pursuit of happiness - fiat lux. So mote and blessed be. *Veritas omnia vincit.* Truth conquers all.

[144] Rev 2:26-29.

Appendices

APPENDIX A

Pi & Phi Encoded in Alphabet

William Eisen in "The Universal Language of Cabalah" describes that, in the generation following the Norman Conquest, the English alphabet was willfully created by adopting the Roman alphabet and adding three new letters to it - JUW. (p.21, 77-79) As a result the Pi and Phi ratios are encoded, symbolically encoding a pentacle. See the chapter "English Alphabet as Pentacle" for more information. This appendix details how the ratios may be read.

You may reveal the encoded Pi and Phi ratios by arranging the characters in a circle and separating the symmetrical characters from the unsymmetrical ones. In the English alphabet, the value of Pi (3.1416) is encoded in the counts of the unsymmetrical characters and can be obtained from the diagram by starting with JKL for the number three and proceeding counter clockwise. The decimal point is encoded in the next symmetrical character M. The one is the single character sequence of N separated from the next by the symmetrical letter O. The four is contained in the sequence PQRS and separated from the next number by the symmetrical TUVWXY. The next number one is the Z followed by A. The final number six is the BCDEFG and terminated by the symmetrical HI.

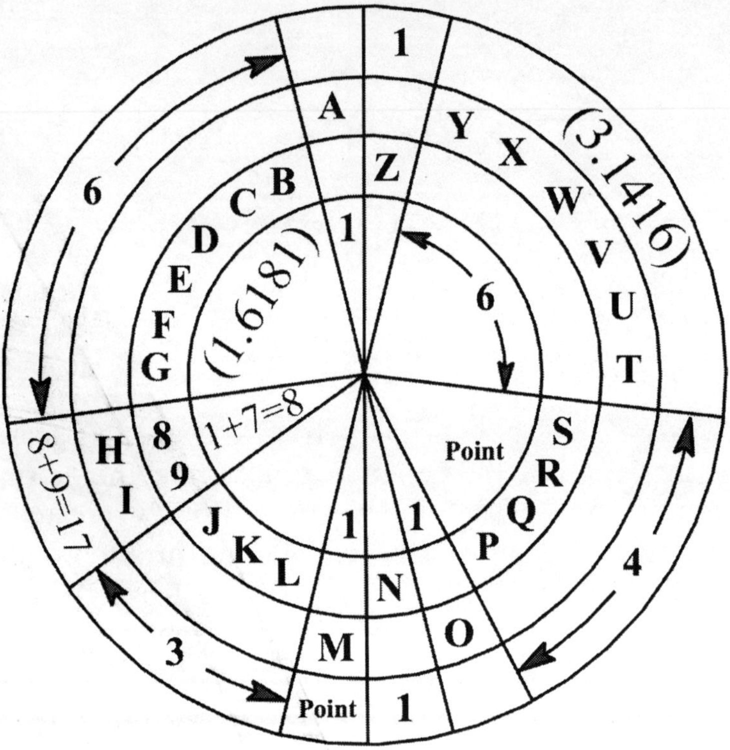

Figure 13 Cipher of English Alphabet

The value of Phi is encoded in the symmetrical characters. The value of Phi, which is 1.6181, can be obtained from the diagram by starting with O for the number one and proceeding counter clockwise. The decimal point is encoded in the unsymmetrical characters PQRS. The six is the TUVWXY and separated from the next number by the Z. The next one is the A followed by the unsymmetrical BCDEFG. The eight is cleverly encoded in the HI characters and separated from the next value by JKL. The value of eight is obtained by adding the numerical position values of H and I and then reducing numerically. Thus, H (8) plus I (9) is 17, and 1+7 is 8. The final number one is the M and is terminated by the N.

This information may also be found in, "The Universal Language of Cabala," by William Eisen (1989).

APPENDIX B

Astronomical Data

Starting 6/11.4/2000 day increment = 2920.00 ct = 1
2920 days are 8*365 days or 8 earth years.
Every 2920 days earth and Venus find themselves
in approximately the same positions.

0) Elapsed days = 0.00
Earth longitude = 90.00, latitude = -0.00, radius = 1.02 A.U.
Venus longitude = 269.99, latitude = 0.24, radius = 0.72 A.U.

1) Elapsed days = 2920.00
Earth longitude = 88.15, latitude = -0.00, radius = 1.02 A.U.
Venus longitude = 268.31, latitude = 0.14, radius = 0.72 A.U.

Starting 6/11.4/2000 day increment = 1460.00 ct = 2
1460 days are 4 years, this is the time elapsed
from superior conjunction to inferior conjunction
of Venus with earth with earth in same location.

0) Elapsed days = 0.00
Earth longitude = 90.00, latitude = -0.00, radius = 1.02 A.U.
Venus longitude = 269.99, latitude = 0.24, radius = 0.72 A.U.

1) Elapsed days = 1460.00
Earth longitude = 89.07, latitude = 0.00, radius = 1.02 A.U.
Venus longitude = 90.36, latitude = -0.26, radius = 0.73 A.U.

2) Elapsed days = 2920.00
Earth longitude = 88.15, latitude = -0.00, radius = 1.02 A.U.
Venus longitude = 268.31, latitude = 0.14, radius = 0.72 A.U.

Starting 6/11.4/2000 day increment = 730.00 ct = 4
730 days are 2 years, every two years when earth is back in its same position, Venus will be found 90 degrees counter clockwise from its last position. Connecting the positions in order will trace a square.

0) Elapsed days = 0.00
Earth longitude = 90.00, latitude = -0.00, radius = 1.02 A.U.
Venus longitude = 269.99, latitude = 0.24, radius = 0.72 A.U.

1) Elapsed days = 730.00
Earth longitude = 89.53, latitude = -0.00, radius = 1.02 A.U.
Venus longitude = 0.69, latitude = 3.38, radius = 0.72 A.U.

2) Elapsed days = 1460.00
Earth longitude = 89.07, latitude = 0.00, radius = 1.02 A.U.
Venus longitude = 90.36, latitude = -0.26, radius = 0.73 A.U.

3) Elapsed days = 2190.00
Earth longitude = 88.61, latitude = 0.00, radius = 1.02 A.U.
Venus longitude = 178.87, latitude = -3.39, radius = 0.73 A.U.

4) Elapsed days = 2920.00
Earth longitude = 88.15, latitude = -0.00, radius = 1.02 A.U.
Venus longitude = 268.31, latitude = 0.14, radius = 0.72 A.U.

Starting 6/11.4/2000 day increment = 584.00 ct = 5
584 days are 584/365=>1.6 earth years and 584/224.6=>2.6 Venus years .3 * 360 degrees == 108
108 degrees are a component of the pentagram.
.6 * 360 == 216 216/3=>72, so for BOTH Venus and Earth the fraction .6 year for their respective year draws a pentagram with each arm drawn as though traced by hand ie. jumping past three points of the star for each 584 day period. Is it cooincidence that the orbital radius of Venus is .72 that of earth?

0) Elapsed days = 0.00
Earth longitude = 90.00, latitude = -0.00, radius = 1.02 A.U.
Venus longitude = 269.99, latitude = 0.24, radius = 0.72 A.U.

1) Elapsed days = 584.00
Earth longitude = 305.31, latitude = 0.00, radius = 0.98 A.U.
Venus longitude = 126.41, latitude = -2.21, radius = 0.73 A.U.

2) Elapsed days = 1168.00
Earth longitude = 159.09, latitude = -0.00, radius = 1.01 A.U.
Venus longitude = 342.19, latitude = 3.30, radius = 0.72 A.U.

3) Elapsed days = 1752.00
Earth longitude = 18.03, latitude = 0.00, radius = 1.00 A.U.
Venus longitude = 196.99, latitude = -3.17, radius = 0.73 A.U.

4) Elapsed days = 2336.00
Earth longitude = 230.06, latitude = -0.00, radius = 0.99 A.U.
Venus longitude = 54.08, latitude = 1.79, radius = 0.72 A.U.

5) Elapsed days = 2920.00
Earth longitude = 88.15, latitude = -0.00, radius = 1.02 A.U.
Venus longitude = 268.31, latitude = 0.14, radius = 0.72 A.U.

Starting 6/11.4/2000 day increment = 365.00 ct = 8
Using 365 earth days as the period places earth always in the same position. 365/224.6 => 1.625. and .625 is 5 eights rotation .625/.125=>5 360*.625->225 degrees 45 degrees is 1/8 circle 5*45=>225. All of these points traced in sequence draws a precise 8 pointed star.

0) Elapsed days = 0.00
Earth longitude = 90.00, latitude = -0.00, radius = 1.02 A.U.
Venus longitude = 269.99, latitude = 0.24, radius = 0.72 A.U.

1) Elapsed days = 365.00
Earth longitude = 89.76, latitude = 0.00, radius = 1.02 A.U.
Venus longitude = 135.41, latitude = -2.58, radius = 0.73 A.U.

2) Elapsed days = 730.00
Earth longitude = 89.53, latitude = -0.00, radius = 1.02 A.U.
Venus longitude = 0.69, latitude = 3.38, radius = 0.72 A.U.

3) Elapsed days = 1095.00
Earth longitude = 89.31, latitude = -0.00, radius = 1.02 A.U.
Venus longitude = 224.25, latitude = -2.26, radius = 0.72 A.U.

4) Elapsed days = 1460.00
Earth longitude = 89.07, latitude = 0.00, radius = 1.02 A.U.
Venus longitude = 90.36, latitude = -0.26, radius = 0.73 A.U.

5) Elapsed days = 1825.00
Earth longitude = 88.84, latitude = -0.00, radius = 1.02 A.U.
Venus longitude = 314.42, latitude = 2.54, radius = 0.72 A.U.

6) Elapsed days = 2190.00
Earth longitude = 88.61, latitude = 0.00, radius = 1.02 A.U.
Venus longitude = 178.87, latitude = -3.39, radius = 0.73 A.U.

7) Elapsed days = 2555.00
Earth longitude = 88.38, latitude = -0.00, radius = 1.02 A.U.
Venus longitude = 44.96, latitude = 2.23, radius = 0.72 A.U.

8) Elapsed days = 2920.00
Earth longitude = 88.15, latitude = -0.00, radius = 1.02 A.U.
Venus longitude = 268.31, latitude = 0.14, radius = 0.72 A.U.

APPENDIX C

Basic Orrery Program

This Appendix shows a computer program written to generate the Earth and Venus conjunctions based upon the text, "Astronomical Algorithms" by Jean Meeus. To calculate planetary positions Meeus describes the planetary theory of P. Bretagnon of the Bureau of Longitudes in Paris, France. This was first published in 1982 as theory VSOP82. The acronym VSOP means "Variations Seculaires des Orbits Planetaires." This rather simple program is based upon rather complicated math, but these algorithms can calculate the planetary positions of Venus and Earth in a three dimensional coordinate system with accuracy 6000 years into the past or future.

The Orrery program is written by the author and may be used freely. Algorithms and equations and VSOP87 planetary tables are from, *Astronomical Algorithms, Second Edition* by Jean Meeus (1998), and are reproduced with permission from the publisher Willmann-Bell Inc., Richmond Virginia. The tables were copied by hand by the author and as this is difficult and error prone it may be prudent to obtain a digital copy of the tables elsewhere.

References:

Jean Meuss, *Astronomical Algorithms, Second Edition* 1988, published by Willmann-Bell, Inc., Richmond, Virginia, USA. Chapter 32, p.217-221 and Appendix III Planets: Periodic Terms, p.416-421.

P. Bretagnon, *Theorie du mouvement de l'ensemble des planetes. Solution VSOP82*, Astronomy and Astrophysics, Vol. 114, pages 278-288 (1982).

P. Bretagnon, G. Francou, *Planetary theories in rectangular and spherical variables. VSOP87 solutions*, Astronomy and Astrophysics, Vol. 202, pages 309-315 (1988).

```
// OrreryBasic.cpp :
//              Visual C++ 6.0 Application based upon algorithms from
//              "Astronomical Algorithms" by Jean Meusse
//              This creates a data.txt file with earth and venus
//              harmonic conjunctions.

#include "stdafx.h"
#include <math.h>
#include <stdio.h>
#include <stdlib.h>
#include "astro.h"
#include "VSOP.h"
char buf[10];


int main(char *argv[], int argc)
{
  if ( testDateFunctions() == 0 ){
    printf("\nFailed algorithm verification.\n\n");
    printf("\nHit return key to terminate.");
    gets(buf);
    exit(0);
    }

  FILE *fp = fopen("data.txt","w");
  if ( !fp ) {
    printf("\nfailed to open data.txt - exiting!");
    exit(0);
    }
```

```
  displayDates(  1, 2920.0, fp );
  displayDates(  2, 1460.0, fp );
  displayDates(  4,  730.0, fp );
  displayDates(  5,  584.0, fp );
  displayDates(  8,  365.0, fp );
  displayDates( 10,  292.0, fp );
  displayDates( 12,  243.0, fp ); // venus sidereal rotation
  displayDates( 20,  146.0, fp );
  displayDates( 40,   73.0, fp );

  fclose(fp);

  printf("\n\nOutput  is data.txt.\n\n");
  printf("\nHit  return key to terminate.");
  gets(buf);

  return 1;
}


void  displayDates( int ctFactor, double  dayFactor, FILE *fp )
{
  double eLongitude, vLongitude;
  double latitude, radius;
  double T;
  Month  month;
  double day;
  int   year;

  month = Jun; day = 11.4; year = 2000;

  double jde = getJDE(month, day, year);
  double jdesave = jde;
  int err;

  err = fprintf(fp,"\nStarting  6/11.4/2000 day increment = %.2f ct = %d",
        dayFactor, ctFactor);

  double venusRot = 90.0;

  switch (ctFactor) {
    case 1:
      fprintf(fp,"\n\n2920  days are 8*365 days or 8 earth years.");
      fprintf(fp,"\nEvery 2920 days earth and venus find themselves ");
      fprintf(fp,"\nin  approximately the same positions.");
      break;
    case 2:
      fprintf(fp,"\n\n1460  days are 4 years, this is the time elapsed");
      fprintf(fp,"\nfrom  superior conjunction  to inferior conjuction");
      fprintf(fp,"\nof  venus with earth with earth in same location.");
      break;
    case 4:
```

```
    fprintf(fp,"\n\n730  days are 2 years, every two years when earth");
    fprintf(fp,"\nis  back in its same position, venus will be found");
    fprintf(fp,"\n90  degrees counter clockwise  from its last position.");
    fprintf(fp,"\nconnecting  the positions  in order will trace a square.");
    break;
  case  5:
    fprintf(fp,"\n\n584  days are 584/365=>1.6  earth years and ");
    fprintf(fp,"\n584/224.6=>2.6  venus  years .3 * 360 degrees == 108");
    fprintf(fp,"\n108  degrees are a component  of the pentagram.");
    fprintf(fp,"\n.6  * 360 == 216 216/3=>72, so for BOTH venus and");
    fprintf(fp,"\nearth  the fraction .6 year for their respective");
    fprintf(fp,"\nyear  draws a pentagram  with each arm drawn as");
    fprintf(fp,"\nthough  traced by hand ie. jumping past three points");
    fprintf(fp,"\nof  the star for each 584 day period. Is it cooincidence");
    fprintf(fp,"\nthat  the orbital radius of venus is .72 that of earth?");
    break;
  case  8:
    fprintf(fp,"\nUsing  365 earth days as the period places earth always");
    fprintf(fp,"\nin  the same position.  365/224.6 => 1.625. and .625");
    fprintf(fp,"\nis  5 eight rotation .625/.125=>5 360*.625=>225 deg");
    fprintf(fp,"\n45  degrees is 1/8 circle 5*45=>225. All of these points");
    fprintf(fp,"\ntraced  in sequence draws a precise 8 pointed  star.");
    break;
  case 10:
    fprintf(fp,"\nUsing  292 earth days 0.8*360=>288  degrees");
    fprintf(fp,"\nearth  rotation. 360-288=>72  deg short of a full turn.");
    fprintf(fp,"\nThus  tracing each point  draws a pentagon in");
    fprintf(fp,"\nclockwise  fashion around the earth. This pentagon");
    fprintf(fp,"\nis  drawn twice in eight years thus specifying  the 4 year");
    fprintf(fp,"\ncycle  again used for Olympic games in ancient  Greece.");
    fprintf(fp,"\n");
    fprintf(fp,"\n292/224.6=>1.3  venus rotations. .3*360=>108 degrees");
    fprintf(fp,"\n108/72=>1.5  thus each point  on venus's circle is 1.5");
    fprintf(fp,"\ntimes  the angle  of the pentagram. ");
    fprintf(fp,"\n108*3=>324  and 360-324=>36");
    fprintf(fp,"\nTracing  the points  over 8 years draws a ");
    fprintf(fp,"\nten  pointed  star counter  clockwise. ");
    fprintf(fp,"\n");
    break;
  case 12:
    fprintf(fp,"\n243  is the sidereal rotation of Venus  (relative to stars)");
    fprintf(fp,"\n2920/243  => 12.016");
    fprintf(fp,"\n");
    break;
  case 20:
    fprintf(fp,"\n\nUsing  146 earth days 146/365=>.4, .4*360=>144 deg");
    fprintf(fp,"\n2*72=>144,  so plotting  the position  of earth will trace");
    fprintf(fp,"\na  pentagram  counter  clockwise  5*146=>730  days or 2");
    fprintf(fp,"\nearth  years, 4 are drawn over the 8 year period.");
    fprintf(fp,"\n");
    break;
  case 40:
```

```
    fprintf(fp,"\n\nUsing 73 earth days 73/365=>.2, .2*360=>72 degrees");
    fprintf(fp,"\nrotational earth orbit. Thus in one year 73*5=>365");
    fprintf(fp,"\ntracing the points will precisely a pentagon around");
    fprintf(fp,"\nthe pentagram drawn using the 146 day cycle. 40/5=>8");
    fprintf(fp,"\nthus 8 pentagons are traced in 8 years.");
    fprintf(fp,"\n");
    fprintf(fp,"\n73/224.6=>.325, .325*360=>117 degrees 117*3=>351");
    fprintf(fp,"\n360-351=>9, 360/9=>40 every three periods rests 9");
    fprintf(fp,"\ndegrees short of three points back. Eventually drawing");
    fprintf(fp,"\n40 points around venus.");
    fprintf(fp,"\n");
    break;
  }

int flag=0;
double lastEarth, lastVenus;
if (ctFactor == 1) {
  flag=1;
  }
double adjust1, adjust2;

for ( int i=0; i<=ctFactor; i++) {

  if ( i ) jde += dayFactor;
  err = fprintf(fp,"\n\n%d) Elapsed days = %.2f", i, (jde - jdesave) );

  if ( err < 0 ) {
    printf("\nError writing to file - aborting.");
    printf("\nHit return key to terminate.");
    gets(buf);
    exit(0);
    }

  T = getTimeFromJDE(jde);

  lastEarth = eLongitude;
  getCoordinatesFor(Earth, T, eLongitude, latitude, radius);

  // This works as is because we always start with same date.
  if (i==0) {
    adjust1 = eLongitude - 90;
    adjust2 = 360 - adjust1;
    }

  if ( eLongitude > adjust1 ) eLongitude -= adjust1;
  else eLongitude += adjust2;

  err = fprintf(fp,"\nEarth longitude = %.2f, lat = %.2f, radius = %.2f A.U.",
        eLongitude, latitude, radius);

  lastVenus = vLongitude;
  getCoordinatesFor(Venus, T, vLongitude, latitude, radius);
```

```
    if ( vLongitude  > adjust1 ) vLongitude  -= adjust1;
    else vLongitude  += adjust2;

    err = fprintf(fp,"\nVenus  longitude  = %.2f, lat = %.2f, radius = %.2f A.U.",
          vLongitude,  latitude, radius);
    }
}


void  getCoordinatesFor(
              Planet planet,
              double atTime,
              double &longitude, double &latitude,
              double &radius)
{
  double L, L0, L1, L2, L3, L4, L5;
  double B, B0, B1, B2, B3, B4, B5;
  double R, R0, R1, R2, R3, R4, R5;

  Term *pL0, *pL1, *pL2, *pL3, *pL4, *pL5;
  Term *pB0, *pB1, *pB2, *pB3, *pB4, *pB5;
  Term *pR0, *pR1, *pR2, *pR3, *pR4, *pR5;

  switch ( planet ) {
    case  Venus:
      pL0 = vL0; pL1 = vL1; pL2 = vL2; pL3 = vL3; pL4 = vL4; pL5 = vL5;
      pB0 = vB0; pB1 = vB1; pB2 = vB2; pB3 = vB3; pB4 = vB4; pB5 = 0;
      pR0 = vR0; pR1 = vR1; pR2 = vR2; pR3 = vR3; pR4 = vR4; pR5 = 0;
      break;
    case  Earth:
      pL0 = eL0; pL1 = eL1; pL2 = eL2; pL3 = eL3; pL4 = eL4; pL5 = eL5;
      pB0 = eB0; pB1 = eB1; pB2 = 0;   pB3 = 0;   pB4 = 0;   pB5 = 0;
      pR0 = eR0; pR1 = eR1; pR2 = eR2; pR3 = eR3; pR4 = eR4; pR5 = 0;
      break;
    default: printf("\nUnknown  planet - exiting!"); exit(0);
    }

  // Begin Longitude
  L0 = sumTerms( pL0, atTime );
  L1 = sumTerms( pL1, atTime );
  L2 = sumTerms( pL2, atTime );
  L3 = sumTerms( pL3, atTime );
  L4 = sumTerms( pL4, atTime );
  L5 = sumTerms( pL5, atTime );

  L  = L0;
  L += L1 * atTime;
  L += L2 * pow(atTime,2);
  L += L3 * pow(atTime,3);
  L += L4 * pow(atTime,4);
  L += L5 * pow(atTime,5);
```

```
    L = L / pow(10,8);

    L = convertRadiansToDegrees( L );
    longitude = L;
    // End Longitude

    // Begin Latitude
    B0 = sumTerms( pB0, atTime );
    B1 = sumTerms( pB1, atTime );
    if (pB2) B2 = sumTerms( pB2, atTime );
    if (pB3) B3 = sumTerms( pB3, atTime );
    if (pB4) B4 = sumTerms( pB4, atTime );
    if (pB5) B5 = sumTerms( pB5, atTime );

    B = B0;
    B += B1 * atTime;
    if (pB2) B += B2 * pow(atTime,2);
    if (pB3) B += B3 * pow(atTime,3);
    if (pB4) B += B4 * pow(atTime,4);
    if (pB5) B += B5 * pow(atTime,5);
    B = B / pow(10,8);

    B = convertRadiansToDegrees( B );
    latitude = B;
    // End Latitude

    // Begin Radius
    R0 = sumTerms( pR0, atTime );
    R1 = sumTerms( pR1, atTime );
    R2 = sumTerms( pR2, atTime );
    R3 = sumTerms( pR3, atTime );
    if (pR4) R4 = sumTerms( pR4, atTime );
    if (pR5) R5 = sumTerms( pR5, atTime );

    R = R0;
    R += R1 * atTime;
    R += R2 * pow(atTime,2);
    R += R3 * pow(atTime,3);
    if (pR4) R += R4 * pow(atTime,4);
    if (pR5) R += R5 * pow(atTime,5);
    R = R / pow(10,8);
    radius = R; // Radius is in Astronomical Units.
    // End Radius
}


double getJDE(int month, double D, int Y)
{
    int M = month;
    double A,B, JD;

    // if month is > 2 leave year and month unchanged
```

```
  // if month is 1 or 2 replace year with year - 1, and month = month +12
  // in other words, if Jan or Feb - it is month 13th or 14th of
  // previous year. Meeus p.61.

  if ( M == 1 || M == 2 ){
    --Y;
    M += 12;
    }

  // calculate gregorian calenday
  A = floor(Y/100);
  B = 2 - A + floor(A / 4);

  // set B to zero if this is a Julian date (ie. < Oct 4 1582)
  if ( Y <= 1582 ) {

    if ( Y == 1582 ) {
      if ( M < 10 ) B = 0;
      else{
        if ( M > 10 ) B=0;
        else{ // Month is October
          if ( D > 4 ) B = 0;
          }
        }
      }
    else{
      B = 0;
      }

    }

  JD = floor(365.25 * (Y + 4716)) + floor(30.6001 * (M+1)) + D + B - 1524.5;
  return JD;
}


void  getCalendarDateFromJDE(int &month, double &day, int &year,
double jde)
{
  double Z,F,A,a,C,D,E,B;

  // Astronomical algorithms p.63
  Z = floor(jde + 0.5);
  F = (jde + 0.5) - Z;

  if ( Z < 2299161 ) {
    A = Z;
    }
  else{
    a = floor( (Z - 1867216.25) / 36524.25 );
    A = Z + 1 + a - floor( a / 4 );
    }
```

```
    B = A + 1524;
    C = floor( (B - 122.1) / 365.25 );
    D = floor( 365.25 * C );
    E = floor( (B - D) / 30.6001 );
    day = B - D - floor( 30.6001 * E ) + F;
    if ( E < 14 )
      month = (int)E - 1;
    else
      month = (int)E - 13;

    if ( month > 2 )
      year  = (int)C - 4716;
    else
      year  = (int)C - 4715;

}


int testDateFunctions(void)
{
    // p.64
    if ( !subTestDateFunctions(  1957, 10,  4.81, 2436116.31) ) return 0;
    if ( !subTestDateFunctions(   333,  1, 27.5,  1842713.0) ) return 0;
    if ( !subTestDateFunctions(  -584,  5, 28.63, 1507900.13)) return 0;

    // p.62
    if ( !subTestDateFunctions(  2000,  1,  1.5,  2451545.0) ) return 0;
    if ( !subTestDateFunctions(  1999,  1,  1.0,  2451179.5) ) return 0;
    if ( !subTestDateFunctions(  1987,  1, 27.0,  2446822.5) ) return 0;
    if ( !subTestDateFunctions(  1987,  6, 19.5,  2446966.0) ) return 0;
    if ( !subTestDateFunctions(  1988,  1, 27.0,  2447187.5) ) return 0;
    if ( !subTestDateFunctions(  1988,  6, 19.5,  2447332.0) ) return 0;
    if ( !subTestDateFunctions(  1900,  1,  1.0,  2415020.5) ) return 0;
    if ( !subTestDateFunctions(  1600,  1,  1.0,  2305447.5) ) return 0;
    if ( !subTestDateFunctions(  1600, 12, 31.0,  2305812.5) ) return 0;
    if ( !subTestDateFunctions(   837,  4, 10.3,  2026871.8) ) return 0;
    if ( !subTestDateFunctions(  -123, 12, 31.0,  1676496.5) ) return 0;
    if ( !subTestDateFunctions(  -122,  1,  1.0,  1676497.5) ) return 0;
    if ( !subTestDateFunctions(-1000,   7, 12.5,  1356001.0) ) return 0;
    if ( !subTestDateFunctions(-1000,   2, 29.0,  1355866.5) ) return 0;
    if ( !subTestDateFunctions(-1001,   8, 17.9,  1355671.4) ) return 0;
    if ( !subTestDateFunctions(-4712,   1,  1.5,        0.0) ) return 0;
    return 1;

}


int subTestDateFunctions(int  year, int month, double day, double jd)
{
  double day2, jd2;
  int month2, year2;
```

```
  jd2 = getJDE(month, day, year);

  if ( jd2 != jd ) {
    printf("Failed testDateFunctions( getJDE() )\n");
    return 0;
    }

  getCalendarDateFromJDE(month2, day2, year2, jd2);

  if ( (month2 != month) || (year2 != year) ) {
    printf("Failed testDateFunctions( getCalendarDateFromJDE() )\n");
    return 0;
    }

  if ( day != day2 ) {
    // the days can be off slightly
    double tmp;
    if ( day2 > day )
      tmp = day2 - day;
    else
      tmp = day - day2;
    if (tmp > 0.001) {
      printf("Failed testDateFunctions( day != day2 )\n");
      return 0;
      }

    }

  return 1;
}


double getTimeFromJDE(double jde)
{
  // p.163
  return (jde - 2451545.0) / 365250;

}


double sumTerms( Term term[], double T )
{
  int i=1;
  double ltmp = 0;

  while ( term[i].A ) {
    ltmp += term[i].A * cos(term[i].B + term[i].C * T);
    i++;
    }

  return ltmp;
```

```
}


double convertRadiansToDegrees( double radians )
{
  double toRadian = PI/180;
  double ltmp = (radians / toRadian); // convert radians to degrees

  if ( ltmp < 0 ) {
    while (ltmp < 0) ltmp += 360;
    double base = floor(ltmp);
    double rem = ltmp - base;
    ltmp = base + rem;
    }

  else if ( ltmp > 360 ) {
    while (ltmp > 360) ltmp -= 360;
    }

  if ( ltmp > 355 )
     ltmp -= 360;
  return ltmp;
}
```

APPENDIX D

VSOP 87 Tables

```
// File: VSOP.h
// Venus L0
Term vL0[] =
{
    { 0, 0, 0 },
    { 317614667, 0, 0 },
    { 1353968, 5.5931332, 10213.2855462 },
    { 89892, 5.30650, 20426.57109 },
    { 5477, 4.4163, 7860.4194 },
    { 3456, 2.6996, 11790.6291 },
    { 2372, 2.9938, 3930.2097 },
    { 1664, 4.2502, 1577.3435 },
    { 1438, 4.1575, 9683.5946 },
    { 1317, 5.1867, 26.2983 },
    { 1201, 6.1536, 30639.8566 },
    { 769, 0.816, 9437.763 },
    { 761, 1.950, 529.69 },
    { 708, 1.065, 775.523 },
    { 585, 3.998, 191.448 },
    { 500, 4.123, 15720.839 },
    { 429, 3.586, 19367.189 },
    { 327, 5.677, 5507.553 },
    { 326, 4.591, 10404.734 },
    { 232, 3.163, 9153.904 },
    { 180, 4.653, 1109.379 },
    { 155, 5.570, 19651.048 },
    { 128, 4.226, 20.775 },
    { 128, 0.962, 5661.332 },
    { 106, 1.537, 801.821 },
    { 0, 0, 0 }
```

```
};
// Venus L1
Term vL1[] =
{
    { 0, 0, 0 },
    {1021352943053, 0, 0 },
    { 95708, 2.46424, 10213.28555 },
    { 14445, 0.51625, 20426.57109 },
    { 213, 1.795, 30639.857 },
    { 174, 2.655, 26.298 },
    { 152, 6.106, 1577.344 },
    { 82, 5.70, 191.45 },
    { 70, 2.68, 9437.76 },
    { 52, 3.60, 775.52 },
    { 38, 1.03, 529.69 },
    { 30, 1.25, 5507.55 },
    { 25, 6.11, 10404.73 },
    { 0, 0, 0 }
};
// Venus L2
Term vL2[] =
{
    { 0, 0, 0 },
    { 54127, 0, 0 },
    { 3891, 0.3451, 10213.2855 },
    { 1338, 2.0201, 20426.5711 },
    { 24, 2.05, 26.03 },
    { 19, 3.54, 30639.86 },
    { 10, 3.97, 775.52 },
    { 7, 1.52, 1577.34 },
    { 6, 1.00, 191.45 },
    { 0, 0, 0 }
};
// Venus L3
Term vL3[] =
{
    { 0, 0, 0 },
    { 136, 4.804, 10213.286 },
    { 78, 3.67, 20426.57 },
    { 26, 0, 0 },
    { 0, 0, 0 }
};
// Venus L4
Term vL4[] =
{
    { 0, 0, 0 },
    { 114, 3.1416, 0 },
    { 3, 5.21, 20426.57 },
    { 2, 2.51, 10213.29 },
    { 0, 0, 0 }
};
// Venus L5
```

```
Term vL5[] =
{
    { 0, 0, 0 },
    { 1, 3.14, 0 },
    { 0, 0, 0 }
};
//------------------------------------------------------
// Venus B0
Term vB0[] =
{
    { 0, 0, 0 },
    { 5923638, 0.2670278, 10213.2855462 },
    { 40108, 1.14737, 20426.57109 },
    { 32815, 3.14159, 0 },
    { 1011, 1.0895, 30639.8566 },
    { 149, 6.254, 18073.705 },
    { 138, 0.860, 1577.344 },
    { 130, 3.672, 9437.763 },
    { 120, 3.705, 2352.866 },
    { 108, 4.539, 22003.915 },
    { 0, 0, 0 }
};
// Venus B1
Term vB1[] =
{
    { 0, 0, 0 },
    { 513348, 1.803643, 10213.285546 },
    { 4380, 3.3862, 20426.5711 },
    { 199, 0, 0 },
    { 197, 2.530, 30639.857 },
    { 0, 0, 0 }
};
// Venus B2
Term vB2[] =
{
    { 0, 0, 0 },
    { 22378, 3.38509, 10213.28555 },
    { 282, 0, 0 },
    { 173, 5.256, 20426.571 },
    { 27, 3.87, 30639.86 },
    { 0, 0, 0 }
};
// Venus B3
Term vB3[] =
{
    { 0, 0, 0 },
    { 647, 4.992, 10213.286 },
    { 20, 3.14, 0 },
    { 6, 0.77, 20426.57 },
    { 3, 5.44, 30639.86 },
    { 0, 0, 0 }
};
```

```
// Venus B4
Term vB4[] =
{
    { 0, 0, 0 },
    { 14, 0.32, 10213.29 },
    { 0, 0, 0 }
};
//-----------------------------------------------------
// Venus R0
Term vR0[] =
{
    { 0, 0, 0 },
    { 72334821, 0, 0 },
    { 489824, 4.021518, 10213.285546 },
    { 1658, 4.9021, 20426.5711 },
    { 1632, 2.8455, 7860.4194 },
    { 1378, 1.1285, 11790.6291 },
    { 498, 2.587, 9683.595 },
    { 374, 1.423, 3930.210 },
    { 264, 5.529, 9437.763 },
    { 237, 2.551, 15720.839 },
    { 222, 2.013, 19367.189 },
    { 126, 2.728, 1577.344 },
    { 119, 3.020, 10404.734 },
    { 0, 0, 0 }
};
// Venus R1
Term vR1[] =
{
    { 0, 0, 0 },
    { 34551, 0.89199, 10213.28555 },
    { 234, 1.772, 20426.571 },
    { 234, 3.142, 0 },
    { 0, 0, 0 }
};
// Venus R2
Term vR2[] =
{
    { 0, 0, 0 },
    { 1407, 5.0637, 10213.2855 },
    { 16, 5.47, 20426.57 },
    { 13, 0, 0 },
    { 0, 0, 0 }
};
// Venus R3
Term vR3[] =
{
    { 0, 0, 0 },
    { 50, 3.22, 10213.29 },
    { 0, 0, 0 }
};
// Venus R4
```

```
Term vR4[] =
{
    { 0, 0, 0 },
    { 1, 0.92, 10213.29 },
    { 0, 0, 0 }
};
//-------------------------------------------------
// Earth L0
Term cL0[] =
{
    { 0, 0, 0 },
    { 175347046, 0, 0 },
    { 3341656, 4.6692568, 6283.0758500 },
    { 34894, 4.62610, 12566.15170 },
    { 3497, 2.7441, 5753.3849 },
    { 3418, 2.8289, 3.5231 },
    { 3136, 3.6277, 77713.7715 },
    { 2676, 4.4181, 7860.4194 },
    { 2343, 6.1352, 3930.2097 },
    { 1324, 0.7425, 11506.7698 },
    { 1273, 2.0371, 529.6910 },
    { 1199, 1.1096, 1577.3435 },
    { 990, 5.233, 5884.927 },
    { 902, 2.045, 26.298 },
    { 857, 3.508, 398.149 },
    { 780, 1.179, 5223.694 },
    { 753, 2.533, 5507.553 },
    { 505, 4.583, 18849.228 },
    { 492, 4.205, 775.523 },
    { 357, 2.920, 0.067 },
    { 317, 5.849, 11790.629 },
    { 284, 1.899, 796.298 },
    { 271, 0.315, 10977.079 },
    { 243, 0.345, 5486.778 },
    { 206, 4.806, 2544.314 },
    { 205, 1.869, 5573.143 },
    { 202, 2.458, 6069.777 },
    { 156, 0.833, 213.299 },
    { 132, 3.411, 2942.463 },
    { 126, 1.083, 20.775 },
    { 115, 0.645, 0.980 },
    { 103, 0.636, 4694.003 },
    { 102, 0.976, 15720.839 },
    { 102, 4.267, 7.114 },
    { 99, 6.21, 2146.17 },
    { 98, 0.68, 155.42 },
    { 86, 5.98, 161000.69 },
    { 85, 1.30, 6275.96 },
    { 85, 3.67, 71430.70 },
    { 80, 1.81, 17260.15 },
    { 79, 3.04, 12036.46 },
    { 75, 1.76, 5088.63 },
```

```
    { 74, 3.50, 3154.69 },
    { 74, 4.68, 801.82 },
    { 70, 0.83, 9437.76 },
    { 62, 3.98, 8827.39 },
    { 61, 1.82, 7084.90 },
    { 57, 2.78, 6286.60 },
    { 56, 4.39, 14143.50 },
    { 56, 3.47, 6279.55 },
    { 52, 0.19, 12139.55 },
    { 52, 1.33, 1748.02 },
    { 51, 0.28, 5856.48 },
    { 49, 0.49, 1194.45 },
    { 41, 5.37, 8429.25 },
    { 41, 2.40, 19651.05 },
    { 39, 6.17, 10447.39 },
    { 37, 6.04, 10213.29 },
    { 37, 2.57, 1059.38 },
    { 36, 1.71, 2352.87 },
    { 36, 1.78, 6812.77 },
    { 33, 0.59, 17789.85 },
    { 30, 0.44, 83996.85 },
    { 30, 2.74, 1349.87 },
    { 25, 3.16, 4690.48 },
    { 0, 0, 0 }
};
// Earth L1
Term eL1[] =
{
    { 0, 0, 0 },
    { 628331966747, 0, 0 },
    { 206059, 2.678235, 6283.075850 },
    { 4303, 2.6351, 12566.1517 },
    { 425, 1.590, 3.523 },
    { 119, 5.796, 26.298 },
    { 109, 2.966, 1577.344 },
    { 93, 2.59, 18849.23 },
    { 72, 1.14, 529.69 },
    { 68, 1.87, 398.15 },
    { 67, 4.41, 5507.55 },
    { 59, 2.89, 5223.69 },
    { 56, 2.17, 155.42 },
    { 45, 0.40, 796.30 },
    { 36, 0.47, 775.52 },
    { 29, 2.65, 7.11 },
    { 21, 5.34, 0.98 },
    { 19, 1.85, 5486.78 },
    { 19, 4.97, 213.30 },
    { 17, 2.99, 6275.96 },
    { 16, 0.03, 2544.31 },
    { 16, 1.43, 2146.17 },
    { 15, 1.21, 10977.08 },
    { 12, 2.83, 1748.02 },
```

```
    { 12, 3.26, 5088.63 },
    { 12, 5.27, 1194.45 },
    { 12, 2.08, 4694.00 },
    { 11, 0.77, 553.57 },
    { 10, 1.30, 6286.60 },
    { 10, 4.24, 1349.87 },
    { 9, 2.70, 242.73 },
    { 9, 5.64, 951.72 },
    { 8, 5.30, 2352.87 },
    { 6, 2.65, 9437.76 },
    { 6, 4.67, 4690.48 },
    { 0, 0, 0 }
};
// Earth L2
Term eL2[] =
{
    { 0, 0, 0 },
    { 52919, 0, 0 },
    { 8720, 1.0721, 6283.0758 },
    { 309, 0.867, 12566.152 },
    { 27, 0.05, 3.52 },
    { 16, 5.19, 26.30 },
    { 16, 3.68, 155.42 },
    { 10, 0.76, 18849.23 },
    { 9, 2.06, 77713.77 },
    { 7, 0.83, 775.52 },
    { 5, 4.66, 1577.34 },
    { 4, 1.03, 7.11 },
    { 4, 3.44, 5573.14 },
    { 3, 5.14, 796.30 },
    { 3, 6.05, 5507.55 },
    { 3, 1.19, 242.73 },
    { 3, 6.12, 529.69 },
    { 3, 0.31, 398.15 },
    { 3, 2.28, 553.57 },
    { 2, 4.38, 5223.69 },
    { 2, 3.75, 0.98 },
    { 0, 0, 0 }
};
// Earth L3
Term eL3[] =
{
    { 0, 0, 0 },
    { 289, 5.844, 6283.076 },
    { 35, 0, 0 },
    { 17, 5.49, 12566.15 },
    { 3, 5.20, 155.42 },
    { 1, 4.72, 3.52 },
    { 1, 5.30, 18849.23 },
    { 1, 5.97, 242.73 },
    { 0, 0, 0 }
};
```

```
// Earth L4
Term eL4[] =
{
    { 0, 0, 0 },
    { 114, 3.142, 0 },
    { 8, 4.13, 6283.08 },
    { 1, 3.84, 12566.15 },
    { 0, 0, 0 }
};
// Earth L5
Term eL5[] =
{
    { 0, 0, 0 },
    { 1, 3.14, 0 },
    { 0, 0, 0 }
};
// Earth B0
Term eB0[] =
{
    { 0, 0, 0 },
    { 280, 3.199, 84334.662 },
    { 102, 5.422, 5507.553 },
    { 80, 3.88, 5223.69 },
    { 44, 3.70, 2352.87 },
    { 32, 4.00, 1577.34 },
    { 0, 0, 0 }
};
// Earth B1
Term eB1[] =
{
    { 0, 0, 0 },
    { 9, 3.90, 5507.55 },
    { 6, 1.73, 5223.69 },
    { 0, 0, 0 }
};
// Earth R0
Term eR0[] =
{
    { 0, 0, 0 },
    { 100013989, 0, 0 },
    { 1670700, 3.0984635, 6283.0758500 },
    { 13956, 3.05525, 12566.15170 },
    { 3084, 5.1985, 77713.7715 },
    { 1628, 1.1739, 5753.3849 },
    { 1576, 2.8469, 7860.4194 },
    { 925, 5.453, 11506.770 },
    { 542, 4.564, 3930.210 },
    { 472, 3.661, 5884.927 },
    { 346, 0.964, 5507.553 },
    { 329, 5.900, 5223.694 },
    { 307, 0.299, 5573.143 },
    { 243, 4.273, 11790.629 },
```

```
    { 212, 5.847, 1577.344 },
    { 186, 5.022, 10977.079 },
    { 175, 3.012, 18849.228 },
    { 110, 5.055, 5486.778 },
    { 98, 0.89, 6069.78 },
    { 86, 5.69, 15720.84 },
    { 86, 1.27, 161000.69 },
    { 65, 0.27, 17260.15 },
    { 63, 0.92, 529.69 },
    { 57, 2.01, 83996.85 },
    { 56, 5.24, 71430.70 },
    { 49, 3.25, 2544.31 },
    { 47, 2.58, 775.52 },
    { 45, 5.54, 9437.76 },
    { 43, 6.01, 6275.96 },
    { 39, 5.36, 4694.00 },
    { 38, 2.39, 8827.39 },
    { 37, 0.83, 19651.05 },
    { 37, 4.90, 12139.55 },
    { 36, 1.67, 12036.46 },
    { 35, 1.84, 2942.46 },
    { 33, 0.24, 7084.90 },
    { 32, 0.18, 5088.63 },
    { 32, 1.78, 398.15 },
    { 28, 1.21, 6286.60 },
    { 28, 1.90, 6279.55 },
    { 26, 4.59, 10447.39 },
    { 0, 0, 0 }
};
// Earth R1
Term eR1[] =
{
    { 0, 0, 0 },
    { 103019, 1.107490, 6283.075850 },
    { 1721, 1.0644, 12566.1517 },
    { 702, 3.142, 0 },
    { 32, 1.02, 18849.23 },
    { 31, 2.84, 5507.55 },
    { 25, 1.32, 5223.69 },
    { 18, 1.42, 1577.34 },
    { 10, 5.91, 10977.08 },
    { 9, 1.42, 6275.96 },
    { 9, 0.27, 5487.78 },
    { 0, 0, 0 }
};
// Earth R2
Term eR2[] =
{
    { 0, 0, 0 },
    { 4359, 5.6846, 6283.0758 },
    { 124, 5.579, 12566.152 },
    { 12, 3.14, 0 },
```

```
    { 9, 3.63, 77713.77 },
    { 6, 1.87, 5573.14 },
    { 3, 5.47, 18849.23 },
    { 0, 0, 0 }
};
// Earth R3
Term eR3[] =
{
    { 0, 0, 0 },
    { 145, 4.273, 6283.076 },
    { 7, 3.92, 12566.15 },
    { 0, 0, 0 }
};
// Earth R4
Term eR4[] =
{
    { 0, 0, 0 },
    { 4, 2.56, 6283.08 },
    { 0, 0, 0 }
};
```

Bibliography

1. Baigent, Michael. Leigh, Richard. Lincoln, Henry. *Holy Blood, Holy Grail.* New York: Dell Publishing, 1982, 1983.
2. Baigent, Michael. *The Jesus Papers: Exposing the Greatest Cover-Up in History.* New York, NY: HarperCollins, 2006.
3. Barnstone, Willis. *The Other Bible – Ancient Alternative Scriptures.* New York, NY: HarperCollins, 1984.
4. Beckman, Petr. *A History of π.* New York: St. Martins Press, 1971.
5. Blavatsky, Helena P., Gomes, Michael. *Isis Unveiled, Secrets of the Ancient Wisdom Tradition, Madame Blavatsky's First Work: A New Abridgement for Today.* Wheaton Ill.: Theosophical Publishing House, 1972, abridged version 1997.
6. Boron, Robert de., Bryant, Nigel. (translator). *Merlin and the Grail - Joseph of Arimathea, Merlin, Perceval, The Trilogy of Arthurian Romances Attributed to Robert de Boron, Translated by Nigel Bryant.* Cambridge, Great Britain, D.S. Brewer, 2001.

7. Bryant, Nigel. *The High Book of the Grail – A translation of the thirteenth-century romance Perlesvaus.* Cambridge, D.S.Brewer, Revised Edition 2007.
8. Campbell, Joseph. *The Flight of the Wild Gander – Explorations in the Mythological Dimension.* South Bend, Indiana: Regnery/Gateway Inc. 1951, 1969.
9. Campbell, Joseph. *Myths to Live By.* United States of America: The Viking Press, 1972.
10. Capt, E. Raymond, MA, AIA, FSA, Scot. *The Scottish Declaration of Independence:* Muskogee, OK: Artisan Publishers, 1983, 2002.
11. Casper, Max. *Kepler.* New York: Dover Publications, 1959, 1993.
12. Duchane, Sangeet. *Beyond the Da Vinci Code: From the Rose Line to the Bloodline.* Singapore: Barnes & Noble, Inc. 2005.
13. Donahue, William H. *Selections From Kepler's Astronomia Nova.* Santa Fe, New Mexico: Green Lion Press. 2004, 2008.
14. Eisen, William. *The Universal Language of Cabalah.* Marina Del Rey, CA: DeVorss & Company 1989.
15. Eisler, Riane. *The Chalice and the Blade, Our History, Our Future.* New York, NY: HarperCollins 1987.
16. Eschenbach, Wolfram von. Hatto, A. T. (Translator). *Parzival.* London, England: Penguin Books Ltd, 1980.
17. Finkelstein, Israel. Silberman, Neil Asher. *The Bible Unearthed – Archaeology's New Vision of Ancient Israel and the Origin of its Sacred Texts.* New York, NY: Simon & Shuster Inc. 2001.
18. Freke, Timothy. Gandy, Peter. *The Jesus Mysteries, Was the Original Jesus a Pagan God?* New York: Three Rivers Press, 1999.
19. Godwin, Joscelyn. *The Golden Thread, The Ageless Wisdom of the Western Mystery Traditions.* Wheaton IL: Quest Books – Theosophical Publishing House, 2007.

20. Goldsmith, Joel S. *The Mystical I.* New York, Evanston, San Francisco, London: Harper & Row, 1971.
21. Graves, Kersey. *The World's Sixteen Crucified Saviors, or Christianity Before Christ.* Montana, USA: Kessinger Publishing Co.
22. Greene, Brian R. *The Fabric of the Cosmos, Space, Time, and the Texture of Reality,* New York: Vintage Books 2004.
23. Hall, Manly P. *The Secret Teachings of All Ages – An Encyclopedic Outline of Masonic, Hermetic, Qabbalistic, and Rosicrucian Symbolical Philosophy – Being an Interpretation of the Secret Teachings concealed within the Rituals, Allegories and Mysteries of all Ages.* Los Angeles, California: Philosophical Research Society, Inc. 1929, 1988.
24. Hancock, Graham. Bauval, Robert. *Talisman - Gnostics, Freemasons, Revolutionaries, and the 2000-year-old Conspiracy at Work in the World Today.* Hammersmith, London: HarperCollins, 2004
25. Hawking, Stephen. *On the Shoulders of Giants – The Great Works of Physics and Astronomy,* London, Philadelphia: Running Press, 2002.
26. Henry, Matthew. *Matthew Henry's Commentary on the Whole Bible Complete and Unabridged in One Volume* (Peabody, Massachusetts: Hendrickson Publishers Marketing LLC, 1991, 2008).
27. Higgins, Godfrey. *Anacalypsis - an Attempt to Draw Aside the Veil of the Saitic Isis or An Inquiry Into the Origin of Languages Nations and Religions.* Brooklyn, New York: A&B Book Publishers 1836, 1992.
28. Huntley, H.E. *The Divine Proportion - A Study in Mathematical Beauty.* New York: Dover Publications, Inc., 1970.

29. Jean, Roger. *Phylotaxis: A Systematic Study in Plant Morphogenesis.* New York: Cambridge University Press 1994.
30. Knight, Christopher. Butler, Alan. *The Hiram Key Revisited, Freemasonry: A Plan of a New World Order.* New York: Metro Books 2010.
31. Knight, Gareth. *A Practical Guide to Qabalistic Symbolism.* York Beach Maine: Samuel Weiser, Inc. 1965.
32. Liungman, Carl G. *Dictionary of Symbols.* New York, London: W.W. Norton & Company, 1991.
33. Livio, Mario. *The Golden Ratio, The Story of Phi, The World's Most Astonishing Number.* New York: Broadway Books, 2002.
34. Maclean, Kenneth J. M., *A Geometric Analysis of the Platonic Solids and Other Semi-regular Polyhedra – With an introduction to the phi ratio – For Teachers, Researchers, and the Generally Curious*, 2nd Printing, Loving Healing Press, 2015.
35. Martin, Everett Dean. *The Behaviour of Crowds, A Psychological Study.* New York and London: Harper and Brothers Publishers, 1920.
36. McGrath, Alister. *In the Beginning – The Story of the King James Bible and How It Changed a Nation, a Language, and a Culture.* New York: Anchor Books a Division of Random House Inc., 2001.
37. Meeus, Jean. *Astronomical Algorithms, Second Edition.* Richmond Virginia: Willmann-Bell, Inc. 1991, 1998.
38. Murchie, Guy. *The Seven Mysteries of Life, An Exploration of Science and Philosophy.* New York: Houghton Mifflin, 1978.
39. Nietzsche, Friedrich. *The Antichrist: A Criticism of Christianity.* New York, NY: Barnes & Noble, Inc., 1895, 2006.

40. Ouspensky, P.D. *A New Model of the Universe – Principles of the Psychological Method in its Application to Problems of Science, Religion, and Art.* Mineola, NY: Dover Publications 1997, New York: Alfred A. Knopf, Inc. 1931.
41. Pagels, Elaine. *Revelations: Vision, Prophecy, and Politics in the Book of Revelation.* New York, NY USA Viking Penquin, 2012.
42. Pagels, Elaine. *The Gnostic Gospels.* New York: Random House, 1979.
43. Picknet, Lynn. Prince, Clive. *The Templar Revelation, Secret Guardians of the True Identity of Christ.* New York: Touchstone, 1997.
44. Pike, Albert. *Morals and Dogma of the Ancient and Accepted Scottish Rite of Freemasonry.* Montana USA: Kessinger Publishing Company, 1871.
45. Ranke-Heineman, Uta. *Eunuchs For the Kingdom of Heaver – Women, Sexuality, and the Catholic Church.* New York: Doubleday, 1990.
46. Ramacharaka, Yogi. *Mystic Christianity, or the Inner Teachings of the Master.* Chicago, USA: Yogi Publication Society, 1907, 1908.
47. Robinson, John J. *Born in Blood – The Lost Secrets of Freemasonry.* New York, NY: M. Evans & Company, 1989.
48. Rutherford, Adam. *Israel-Britain or Anglo-Saxon Israel – An Explanation of the Origin, Function and Destiny of the Anglo-Celto-Saxon Race in the British Empire and U.S.A.* Muskogee, OK: Artisan Publishers 1934, 2002.
49. Schoch, Robert M. Ph.D. *Forgotten Civilization – The Role of Solar Outbursts in Our Past and Future.* Rochester, Vermont: Inner Traditions 2012.
50. Shipley, Joseph T. *Dictionary of Word Origins.* New York, Dorset Press, MCMXLV.

51. Silk, Roger D., Lintott, James W., [et, al] *Creating a Foundation, The Essential Guide for Donors and Their Advisers.* Princeton, New Jersey USA, Bloomberg Press 2003.
52. Simrock, Karl. *Der Wartburgkrieg.* Cotta, Stuttgart, Augsburg 1858. OCLC 5363166.
53. Smith, Morton Ph.D. *Jesus the Magician.* New York, NY: HarperCollins 1978.
54. Starbird, Margaret. *The Woman With the Alabaster Jar.* Rochester, Vermont: Bear & Company, 1993.
55. Summers, Montague. *The Malleus Maleficarum of Heinrich Kramer and James Sprenger - Translated With Introduction, Bibliography & Notes by the Reverend Montague Summers.* New York: Dover Publications, Inc. 1928, 1948, 1971.
56. Three Initiates. *The Kybalion, A Study of The Hermetic Philosophy of Ancient Egypt and Greece.* Masonic Temple Chicago, Ill.: The Yogi Publication Society, 1912.
57. Twynan, Tracy R. *Soloman's Treasure – The Magic and Mystery of America's Money.* Portland Oregon: Dragon Key Press 2005.
58. Wallace-Murphy, Tim. Hopkins, Marilyn. *Rosslyn - Guardian of the Secrets of the Holy Grail.* Italy: Barnes & Noble, 2000.
59. Waite, Arthur Edward. *The Hidden Church of the Holy Graal – It's Legends and Symbolism Considered in Their Affinity With Certain Mysteries of Initiation and Other Traces of a Secret Tradition in Christian Times.* London: Rebman Limited, 1909.
60. Waite, Arthur Edward. *The Holy Grail – It's Legends and Symbolism.* London: Rider and Co. 1933.

Index

T

V

W

ABOUT THE AUTHOR

The author has spent thirty years working at AT&T Laboratories as a software developer. He has a Bachelor of Arts in Computer Science with a minor in Medieval History from Rutgers University College, class of 1987. He also has a lifelong interest in secret societies, ancient history, and the primordial wisdom traditions. In his spare-time he enjoys writing, travel, scuba diving, aerial yoga, amateur astronomy, and motorcycle restoration. The author is working with his attorney and The Rutgers University Foundation, on the creation of a Literary Trust for the long-term preservation of this book and other writings.

Made in the USA
Monee, IL
15 June 2026

54210781R00146